"This is a refreshing and clear-eyed must-read for all educators that thoughtfully approaches teaching with holistic and collectivist lenses, weaving literacy and life learning with climate stewardship in ways that cultivate students who care about belonging, their communities and the world. Xochitl's ideas provide integrated and grounded practices—our way forward in a challenging, fractured time."

Nawal Qarooni, *literacy consultant, Founder, NQC Literacy, and Adjunct Professor, New York University, USA*

"Xochitl Bentley reimagines climate education by inviting teachers and learners to approach stewardship through storytelling, using six essential stances. This beautiful book includes tangible, classroom-tested strategies that enable students to disrupt overwhelming environmental narratives and craft hopeful counterstories. Transcending disciplinary boundaries, Xochitl advocates for a cross-disciplinary approach to coaching climate literacy. Her attention to intergenerational awareness, kinship, and reciprocity speaks to our shared humanity, igniting our imaginations and calling us to collective action. This refreshing resource empowers educators to ready the next generation of climate stewards to advocate for a more just and sustainable future."

Angela Stockman, *Professional Learning Facilitator and Founder, WNY Education Associates and Make Writing Studios, USA*

Helping Students Become Climate Stewards

This resource examines the way storytelling can play a role in environmental problem-solving and climate stewardship. Narrative not only builds literacy but also fosters students' critical thinking around the ways they inhabit their world. The author examines habits, myths, and mindsets that threaten our planet's ecosystems, and presents "counter-stories" you can use to build your middle and secondary students' capacity for environmental advocacy. Chapters are organized around a framework for developing environmental literacy, each focusing on how storytelling can build the capacity for various roles within the realm of climate stewardship. Via practical entry points and pathways for lesson and unit design, educators can use narrative to help students envision themselves as systems thinkers, communicators, activists, problem solvers, and more. Each chapter uses different kinds of narratives, from fictional parables to comic storyboards, to present practices students can understand and try out. They also include case studies, writing exercises, lesson suggestions, planning tool organizers, and rubrics applicable throughout different curriculum areas. Ideal for any secondary educator grappling with the uncertainties around climate change in their classroom, this resource introduces and encourages inquiry investigation through the power of storytelling to cultivate climate stewardship.

Xochitl Bentley is a national board-certified English teacher at Cleveland Charter High School in California, USA. She is Co-Director of the CSUN Writing Project and a contributing writer at Moving Writers. She is a certified California naturalist, a certified California environmental educator, and a member of the Climate Reality Leadership Corps.

Also Available from Routledge Eye On Education
(www.routledge.com/eyeoneducation)

Teaching Climate Science in the Elementary Classroom: A Place-Based, Hope-Filled Approach to Understanding Earth Systems
Stephanie Sisk-Hilton

Habitats in the United States, Grade K: STEM Road Map for Elementary School
Carla C. Johnson, Janet B. Walton, Erin E. Peters-Burton

Habitats Local and Far Away, Grade 1: STEM Road Map for Elementary School
Carla C. Johnson, Janet B. Walton, Erin E. Peters-Burton

Hydropower Efficiency, Grade 4: STEM Road Map for Elementary School
Carla C. Johnson, Janet B. Walton, Erin E. Peters-Burton

Culturally Responsive and Sustaining Science Teaching: Teacher Research and Investigation from Today's Classrooms
Edited by Elaine V. Howes and Jamie Wallace

Teaching STEAM Through Hands-On Crafts: Real-World Maker Lessons for Grades 3–8
Christine G. Schnittka and Amanda Haynes

The Speed of Green, Grade 8: STEM Road Map for Middle School
Carla C. Johnson, Janet B. Walton, Erin E. Peters-Burton

Place-Based Scientific Inquiry: A Practical Handbook for Teaching Outside
Benjamin Wong Blonder, Ja'Nya Banks, Austin Cruz, Anna Dornhaus, R. Keating Godfrey, Joshua S. Hoskins, Rebecca Lipson, Pacifica Sommers, Christy Stewart, Alan Strauss

Composting, Grade 5: STEM Road Map for Elementary School
Carla C. Johnson, Janet B. Walton, Erin E. Peters-Burton

Global Populations Issues, Grade 7: STEM Road Map for Middle School
Carla C. Johnson, Janet B. Walton, Erin E. Peters-Burton

Helping Students Become Climate Stewards

Storytelling for Environmental Advocacy and Problem-Solving

Xochitl Bentley

Routledge
Taylor & Francis Group
NEW YORK AND LONDON

Designed cover image: © Getty Images

First published 2026
by Routledge
605 Third Avenue, New York, NY 10158

and by Routledge
4 Park Square, Milton Park, Abingdon, Oxon, OX14 4RN

Routledge is an imprint of the Taylor & Francis Group, an informa business

© 2026 Xochitl Bentley

The right of Xochitl Bentley to be identified as author of this work has been asserted in accordance with sections 77 and 78 of the Copyright, Designs and Patents Act 1988.

All rights reserved. No part of this book may be reprinted or reproduced or utilised in any form or by any electronic, mechanical, or other means, now known or hereafter invented, including photocopying and recording, or in any information storage or retrieval system, without permission in writing from the publishers.

Trademark notice: Product or corporate names may be trademarks or registered trademarks, and are used only for identification and explanation without intent to infringe.

ISBN: 978-1-032-85188-4 (hbk)
ISBN: 978-1-032-85187-7 (pbk)
ISBN: 978-1-003-51700-9 (ebk)

DOI: 10.4324/9781003517009

Typeset in Palatino
by Apex CoVantage, LLC

For my husband, Graham.
I'll always be your date at Carla's Country Kitchen.

For my parents.
You showed me that teaching is a magical art.

For my daughter, Josie.
You are my heartbeat.

Contents

Meet the Author . . . . *x*
Preface . . . . *xi*
Acknowledgments . . . . *xiii*

    Introduction 1

1 **Climate Stewards as Communicators** 14

2 **Climate Stewards as Systems Thinkers** 46

3 **Climate Stewards as Scenario Developers** 89

4 **Climate Stewards as Environmental Justice Activists** 105

5 **Climate Stewards as Problem Solvers** 134

6 **SEL and Climate Stewardship** 150

7 **Project-based Learning Design That Fosters Environmental Literacy** 177

Meet the Author

Xochitl Bentley, NBCT, is a high school English teacher at Cleveland Charter High School in the Los Angeles Unified School District. She is a certified California naturalist, a certified California environmental educator, and a climate reality leader. She holds a bachelor's degree in English from the University of San Francisco, a master's degree in English from the University of Wisconsin-Madison, and a master's degree in education from Claremont Graduate University. She began her career teaching composition at the University of Wisconsin-Madison. She has written articles appearing in the *Oregon English Journal*, *California English*, Newsela, the NCTE blog, and *Education Week*. As a regular contributor at Moving Writers, she writes about lesson design focused on climate stewardship. In 2017, she was one of 12 American educators chosen for the Fulbright Japan–U.S. teacher exchange program to learn more about education for sustainable development.

Xochitl is a third-generation teacher. Her childhood memories are awash with images of classroom scenes and scents. She eagerly helped her parents set up their classrooms before the start of each school year, and the sight of all those new school supplies seemed to symbolize the promise of a new beginning. She is passionate about helping teachers find balance and identify communities of care where they feel supported and valued. The experiences she has gained through the CSUN Writing Project make her career choice and writing life feel sustainable and joyful. As a co-director, she loves learning from other teachers and bringing their ideas back to the classroom.

Preface

In 2017, I took part in a teacher exchange program that allowed me to visit Tokyo and Kitakyushu City in Japan. American and Japanese teachers worked in collaborative teams to share cross-cultural ideas for educating about sustainability. As I walked the streets in Tokyo, I noticed there were no trash cans for me to dispose of my paper coffee cup. Similarly, as I toured schools in both cities, I noticed there were no trash cans for collecting litter in the school hallways and bathrooms. Unlike where I live in the United States, where a "throwaway culture" is evident in the teeming, overflowing trash cans seen in public places, these cities encourage reusing materials and reducing waste. It was an early moment in my thinking about the infrastructure decisions cities make to encourage circular economies: where systems are designed to be restorative and regenerative.

This insight led me to consider when students have opportunities to consider the differences between circular and linear economies. As a secondary English teacher, I wondered how I could weave a discussion about urban planning decisions with the skill building typically associated with the English language arts discipline. Up to this point, I had to create my own pathways for teaching about climate justice through seeking out colleagues who also wished to cultivate environmental stewardship and design lessons together. Though I have had many rewarding experiences collaborating with like-minded colleagues on one-off projects, built around activities such as textile recycling events or beach cleanup visits, the recognition that the challenges associated with climate change require sustained, iterative opportunities for students to solve inspired a deeper consideration of the transdisciplinary potential of environmental literacy.

I vividly remember conversations I had with various educators during my visit to Japan: on the bus ride through Asakusa, on the flight to Kitakyushu City, and each morning in the Tokyo hotel breakfast room. Even though it is unlikely I will see the two dozen or so people I met on this trip again in person, snippets of our conversations float through my mind whenever I'm seeking inspiration. Though we came from different places and worked in different teaching contexts, we possessed a common denominator: our intellectual labor and emotional labor were deeply intertwined. We sought to learn how to improve our instruction about sustaining our planet, which means focusing on communication alongside scientific concepts.

This book is about the critical role storytelling plays in climate stewardship and how environmental problem-solving depends on civic action that offers counterstories to our current unsustainable trajectory. The chapters are organized around a framework outlining practices for cultivating climate stewards that all hinge on storytelling: highlighting the skills of communicators, systems thinkers, scenario developers, environmental justice activists, and problem solvers.

Acknowledgments

Thank you to Heather Swan and Terrapin for allowing me to include Heather's poem in my book:

"Pesticide VII: Victor" from *A Kinship with Ash* (Terrapin Books) Copyright © 2020 by Heather Swan. Reprinted by permission of Terrapin Books.

Thank you to Moving Writers for allowing me to include excerpts from articles I have written and an invitation to the best sandbox.

Thank you to Julia Dolinger, Sofia Cohen, and the rest of the Routledge team for bringing this book to life. You guided a first-time author with patience, kindness, and generosity. I'm so grateful.

Thank you to CSUN Writing Project teacher leaders for community that nourishes me. Thank you especially to Jenn Wolfe and Melissa Wood-Glusac for conversations in the Zoom ether during the pandemic.

Thank you to members of the habitat restoration team in San Francisquito Canyon: Alyssa Walker, Matthew Loftis, Hannah Crispi, Sage Cobos, and Massimo De Maria. You gave me a sense of rootedness in the fire scar. Thank you for planting good things.

Thank you to Angela Stockman, Matthew Johnson, Nawal Qarooni, Jennifer Fletcher, and Trevor Aleo for inspiring me. To my mentor Rebekah O'Dell: your example and encouragement have sustained me during life's hardest moments.

Thank you to Heather Zajdel, Erica Schatz, Hillary Michelle, Mary Ann Ng, Myah Lunceford, Vincenzo Loconte, Abigail Lund, Eric Wilson, and Kirstin Bullington for sharing ideas about how we can foster environmental stewardship and engaging in reflection with me. Your students are so, so lucky.

Thank you to former students Alexa Romero, Emma Ortega, Tyler Lee, Kayla Do, Joseph Lee, Austin Buranapan, Alondra Garcia, Maddie Dameshek, Kailo Pascual, Kingsten Ko, Charlie

Perez, Kayla Barrios, Julienne Sta. Maria-Padame, and Yijing Fang for granting me permission to feature their work in these pages. I learn so much from all of you.

Corina and Cassie, thank you for decades of friendship and many laughter-filled phone calls.

Thank you to Vronnie and Luna for lessons in tenacity. You two help me find every sunbeam.

Introduction

Storytelling Is a Survival Skill

The term "climate steward" offers a broad entry point into thinking about how we can protect the imperiled natural world and ourselves. Thinking about problem-solving in terms of climate impact and needed intervention is not just the concern of the scientist or the policymaker or the journalist. Climate stewardship requires the skills of multiple disciplines and invokes the idea of careful and responsible management. It suggests the necessity of developing communities of care that can begin in the classroom.

Nowadays, many students benefit from school assemblies that convey warning messages about the bystander effect: how, in the presence of others, individuals are discouraged from intervening in an emergency situation. The bystander effect directly relates to the psychological phenomenon called "the diffusion of responsibility"—the more people, the less likely anyone will intervene because they assume someone else will take action. However, some researchers have taken issue with this finding, stating there is evidence that indicates people will intervene if they are aware the situation is critical.

When it comes to the climate crisis, most educators are aware of the criticality of our situation. But we often feel we lack the tools to intervene within the constraints of our narrowly defined subject content area or the stated parameters of a given curriculum. Many educators find conversations related to climate change to be too full of "gloom and doom" or believe themselves ill-prepared to discuss related scientific concepts. Educators have a unique opportunity to help others intervene because we have a transdisciplinary tool in our grasp: storytelling. A tool that can be wielded in the manner of a blunt instrument, like a bludgeoning newspaper headline. A tool that can wrest a listener's wandering attention with a gracefully plotted personal vignette. A tool that students can use to communicate this truth: "My story, the one that no one but I can tell, crucially informs my advocacy." Stories can engage awareness and attention when the scale and severity of human effects on the planet seem too overwhelming to contemplate.

I start with the question: "How has this story been told?"

Not "What scares you about climate change?" Or "What makes you angry about climate change?" But "How has this story been framed?"

In emphasizing the profound ecological implications of stories not immediately recognizable as stories, Arran Stibbe explains: "It is through language that the natural world is mentally reduced to objects or resources to be exploited, and it is through language that people can be encouraged to respect and care for the systems that support life" (Stibbe, 2021, p. 2).
Rethinking what Stibbe calls the stories we live by—stories in the minds of multiple people across a culture—means scrutinizing underlying narratives that we "are exposed to without consciously selecting them or necessarily being aware that they are just stories" (p. 5).

Before helping students ask people to commit to eco-conscious behaviors, we need to help them by modeling "reframing"—rethinking the terrain of what we imagine and becoming mindful of the stories we have been taught that lead to "narrative foreclosure." Psychologist Ernst Bohlmeijer and colleagues define narrative foreclosure as "the conviction that no new interpretation of one's past nor new commitments and experiences in one's future are possible that can substantially change one's life story" (2011, p. 364). In many ways, becoming a climate steward means unlearning narratives that do not encourage our long-term flourishing.

My goal is to help educators see what climate stewardship can look like in the actual teaching trenches. My approach is based on classroom-tested strategies that start with the question: How can improving our storytelling help us advance environmental policies that protect us?

We often possess the political will to effect change but struggle with finding meaningful and effective points of intervention. Stories can help us move beyond merely learning facts about climate change. Stories can help us foster a proactive approach to advocating for our liveable futures. Something that complicates our ability to respond to the climate crisis is the "slow violence" characterizing its effect. Rob Nixon (2011) defines slow violence as "a violence that occurs gradually and out of sight, a violence of delayed destruction dispersed across time and space, an attritional violence that is typically not viewed as violence at all" (2). Because we cannot comprehend the scope of ongoing environmental degradation such as deforestation or the acidification of oceans in one glance, stories about the climate crisis are not easily adapted to TV news segments that average 3–4 minutes in length. This definition usefully illuminates one of the major barriers to accessing substantive climate news stories and draws attention to how the impacts of climate change play out across a range of temporal scales.

The barriers posed by the seeming invisibility of many climate change impacts are not the only obstacles to taking action. Our default storytelling habits thwart our abilities to meaningfully respond to and mitigate the climate crisis. We can still mitigate the future we all seem to be hurtling toward if we shape stories that offer counternarratives.

The Skills of Climate Stewardship Rely on Counterstory

When it comes to fostering environmental stewardship, storytelling is essential because it is the only way we can develop counterstories to the stories we already know. In this book, I outline how six storytelling stances can help learners disrupt harmful mindsets that perpetuate the climate crisis. When we identify environmental transgressions, we are not always sure how to

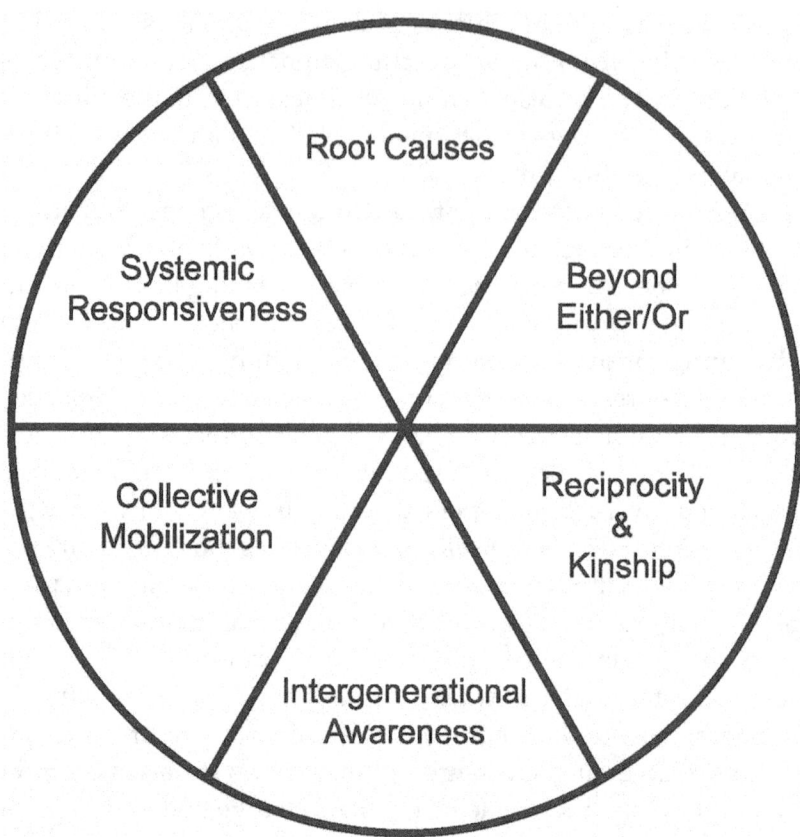

FIGURE I.1 Storytelling Stances

translate our own awareness into lesson design, teachable tools, and collaborative labor. By consciously embedding these stances in our story making, we strengthen our capacity to imagine better futures together.

Storytelling for Systemic Responsiveness

Climate stewards do not only plan with the benefit of hindsight. They plan with foresight. Shifting from a merely reactive posture, in which we take care of ourselves after disaster strikes, means purposefully developing a proactive approach and anticipating what we need to do to protect ourselves and each other. Examples drawn from real-world accounts, as well as from fictional stories, teach us the necessity of shifting from a reactive stance to climate disasters to planning proactive approaches that involve scrutinizing the structures and systems in place and communicating needed improvements. Creating opportunities for our students to assess how systems are both responsive and proactive will help them meet the challenges of both climate change mitigation and adaptation.

This storytelling stance counters the tendency to talk about disaster events or public health failures as one-off, isolated events. Instead, we consider how climate stewardship looks at problems from a structural lens and tackles questions about infrastructure and institutional support.

Storytelling About Root Causes

The symptoms of our climate crisis are often confused with its root causes. For this reason, students need experience and guidance with gathering, sifting through, and synthesizing data sources. Looking at different data points is necessary for making informed decisions, such as recommending how a city or a program can better direct resources. Thinking about data collection in terms of root causes helps us surface those deeply entrenched mindsets and habits that perpetuate the problem.

One way we can help students offer counterstories to shallow narratives that only scratch the surface is to highlight the value of ethnographic work. Tricia Wang (2016) makes an argument for integrating "big data"—quantitative data captured at large scale and "thick data"—qualitative data emerging from ethnographic research methods that uncovers people's stories. By including stories, organizations can glean information about emotions and context that is not easily measured or captured in a report downloaded from a data dashboard. When students take the time to learn from stories, they gather an invaluable data point. They learn about the "sensemaking" efforts of the storyteller and how one story might have an outsized impact due to its pointing to root causes usually obscured by a tidy dataset. They learn that one piece of data is not the whole story.

Storytelling Beyond Either/Or

We must teach our climate stewards to look beyond "either-or" thinking, commonly encountered in arguments that frame voter decision-making as coming down to two options: for example, you can either protect the environment or grow the economy. The Either-Or is a version of the logical fallacy known as the false dilemma. It presents situations with only two outcomes (positive and negative) when in reality there are multiple possibilities. This type of thinker acts as if there are only two mutually exclusive options, thereby leading us to make decisions without knowing all the choices. When problem solvers engage in binary thinking, they often frame the option too narrowly. Instead of being manipulated into thinking there are only two options, we can employ storytelling that invites our audience to imagine other possibilities. For example, we can ask: is it possible to invest in renewable energy in ways that actually boost the economy and spur job creation?

In the book *Both/And Thinking*, Wendy K. Smith and Marianne W. Lewis encourage readers to interrogate this framing:

> What if the problem is that we are too narrow in how we frame the problem? We see two roads and think that those are our only options, rather than diving deeper, asking why we feel forced to make that choice to begin with.
>
> (2022, p. 46)

Instead of seeing climate stewardship as an obstacle to economic growth, we could bolster job training programs in sectors most affected by this transition and better communicate the opportunities existing in "green" careers.

Storytelling for Reciprocity and Kinship

Educators must help students understand that the way we conceptualize the nature–people relationship critically informs our capacity for stewardship. Instead of thinking of nature in terms of its instrumental worth to humans, we can help students focus on kinship. The word "kinship" emphasizes the deep interdependence that exists between humans and the more-than-human world. In the Introduction to the book series *Kinship: Belonging in a*

World of Relations (2021), Gavin Van Horn explains how "kinning" is an exploration of how to rightly relate: "From a kinship perspective, the landscapes of which humans are a part—including rocks, rivers, oceans, prominent geographic features, and other nonhuman plant and animal persons—provide a shared sense of place and require appropriate human care and respect" (p. 6).

Since the 1970s, the field of "deep ecology" has diagnosed Western society's anthropocentrism as humanity's most urgent problem. This environmental philosophy promotes the inherent value of nature rather than its instrumental worth to humans. Robin Wall Kimmerer's writing on reciprocity (2020) can anchor our thinking for helping learners replace harmful ways of thinking based on an extractive relationship with nature. Extractive thinking reduces complex biodiverse relationships to mere human property or resources to be exploited. Educators can foster awareness of counterstories based on systems of reciprocity between living beings, not on systems of extraction. In the book *Braiding Sweetgrass: Indigenous Wisdom, Scientific Knowledge and the Teaching of Plants,* Kimmerer explores her own history of loss and recovery as a member of the Citizen Potawatomi Nation and describes how Indigenous perspectives can transform engagement with a living world. Her writing can inspire our efforts in what she calls the "re-story-ation" of our relationship to the earth: finding ways to move away from domination-oriented relationships to partnerships where mutual flourishing is emphasized.

Box I.1: Key Terms

Kinship: an awareness of ecological interdependence grounded in relationships of belonging

Reciprocity: a mutual exchange or dependence that benefits both, each, and all but also includes mutual responsibilities

Anthropocentrism: the belief that the human being exists at the center of the universe

Extractivism: the practice of extracting natural resources for profit

Storytelling With Intergenerational Awareness

The word "sustainability" is ubiquitous, but it is rarely defined. Taking the time to unpack this concept benefits our ability to weigh what kind of inheritance we leave for future generations. In the UN Brundtland commission report, *Our Common Future* (1987), sustainability is defined as "meeting the needs of the present without compromising the ability of future generations to meet their own needs." This definition highlights the need for cultivating intergenerational awareness as we prepare students to mitigate the harmful human impacts associated with climate change. Rewriting our story with this type of awareness means thinking of "inheritance" in terms of the policies, laws, and regulations we enact. It means thinking of "inheritance" in terms of our own civic engagement and voting habits. It means reflecting on how we can invest in the future by divesting from fossil fuels. Thinking as a future ancestor means remembering to think beyond ourselves with a sense of solution-oriented urgency and modeling this commitment for our students.

Storytelling for Collective Mobilization

We need to replace the myth of rugged individualism with a belief in the power of collective mobilization. For too long, the idea of the self-reliant individual who can succeed on their own, pulled up by their own bootstraps, has dominated the public imagination. A romanticized concept often associated with the American frontier, this myth again reared its head during the coronavirus pandemic. In a 2022 *Scientific American* article, "How a Virus Exposed the Myth of Rugged Individualism," Robin G. Nelson writes "Communities that were already heavily invested in social safety nets with measures such as paid sick leave were able to lower COVID rates. Those invested in the idea of self-sufficiency and individualism prolonged suffering and loss of life."

To help students counter the myth of self-reliance, we need to help them invest in social capital. Social capital, a phrase derived from economics, describes the ways that groups invest in each other. Instead of money, the investment is in the relationships we have with one another. We can take the networks we build in schools into communities and vice versa. By helping students identify ways that they can be active members in networks of support, we foster the ability to develop assets through community partnerships. We can feel heartened by the shared values that arise from these networks and the knowledge that it is easier to address systemic change in partnership.

How This Book Is Organized

Each chapter offers lesson ideas for teaching stories as stimulus texts for inquiry and investigation. Since students need both scientific concepts and communication skills to become effective climate stewards, this book highlights the importance of storytelling in helping students foster awareness and communicate calls to action. In the book *The Story is in Our Bones: How WorldViews and Climate Justice Can Remake a World in Crisis*, Osprey Orielle Lake stresses the necessity of shifting our mental paradigms for the work needed to be done:

> Networks of change-makers, systems thinkers, activists, writers, and artists from every background have recognized the need for systemic change because we know that while installing solar panels, deploying wind turbines, recycling, and implementing energy efficiency are real solutions and absolutely paramount, these alone cannot address the deep layers of societal evolution that this moment demands.
>
> (2024, p. 9)

As a teacher who navigates many external expectations regarding what is happening in my classroom, it is tempting to imagine that leaving the profession and working at an environmental nonprofit might more directly contribute to the fight for our planet: the fight for our liveable future and a habitable, healthy biosphere. I rein in this impulse by imagining the potential multiplying effect of working with students: how students, when given the knowledge and communication tools to advocate on behalf of our climate, model for others how to be engaged climate stewards who can problem-solve.

This train of thought relies on a key verb: imagine. In our daily instruction, we might find ourselves prioritizing other skills, but I want to argue that imagination is a critical survival skill. Without the resources of our imagination, we can't act on our hope to mitigate and even reverse some of the effects of human-caused climate change. This is why I think literacy instruction is a key tool in our fight for survival. To share and exchange ideas for sustaining our planet home, we need stories. We need stories that help us comprehend the root causes of our climate crisis. We need stories that help us rewrite the future we are sure to experience if we do not do everything in our power to lower the global temperature. We need to become storytellers who can advocate on behalf of each other and Nature. Stories offer models of action that create the multiplying effect needed to build environmental stewardship capacity.

Climate reporter Denise Hruby (2024) stresses the potential for journalists to broaden the reach of their news coverage by reporting stories through a climate change lens:

> Every story has a climate change angle: every election story, every economic decision. Even sports: Think about how your local stadium is preparing for, say, flooding? How is your local team going to play during a heat wave, how are the athletes going to stay safe? If you bring climate into the beats you're already covering, you'll not only have a plethora of new stories, but you're going to better inform your readers, and you're going to better serve your community.
>
> (salatainstitute.harvard.edu)

"If you bring climate into the beats you're already covering . . ." Teachers, we can transfer this invitation to our classroom context: if we bring climate into the skill instruction we're already doing, the possibilities skyrocket.

Environmental storytelling transcends disciplinary boundaries. When my students are able to weave lessons from their biology class with lessons from their history class, they exhibit a sense of satisfaction that they can bring the seemingly disparate hours of their learning day together. We can support emerging environmental storytellers by integrating ways of thinking about our connections to the natural world with our focus on literacy instruction—meaningful reading and writing practices are happening in all content areas. As we help students craft narratives with an audience in mind, we can reflect on how to broaden the reach of environmental storytelling and ideas for problem-solving.

When I have collaborated with colleagues on climate stewardship topics or designed lesson materials on my own, working within curricular parameters sometimes made it hard to do so. I know educators lack time to comb through resources, so I hope this book will offer you guidance and clarity about how we can support the climate stewards in our classrooms.

Chapter 1 is devoted to helping students develop and expand their repertoire of "craft moves" for effective communication as climate stewards, offering examples for employing the craft moves of mentor texts that model effective place-based calls to action. Readers will learn to "read like writers" as they identify and practice craft moves for communicating the specific features of place: geography, contexts, and landscapes.

Chapter 2 illuminates why practicing systems thinking helps students distinguish between the symptoms and root causes of environmental problems. Readers will learn how to use causal loop diagramming to elicit ideas about the most effective intervention, as well as how to plan podcasting conversations to share information about this process. These tools will prepare students to weigh solutions with a more nuanced understanding of how problems are perpetuated through systems.

Chapter 3 introduces scenario development strategies that prepare students for future-oriented thinking—how problems might occur or evolve over time. Readers will learn to adopt the lens of environmental urban planning, as well as how to use the VUCA tool for making suggestions for restructuring, allocating resources, and expanding information networks.

Chapter 4 stresses the importance of advocating for environmental justice, which is distinct from a more broadly understood environmentalism. Readers will learn about historic and ongoing disparities in how communities are protected from environmental hazards, as well as the role of comic storyboarding in preparing students for their first steps in environmental justice activism. Educators will feel prepared to identify indicators of environmental hazards, as well as the importance of heeding the warnings of community voices.

Chapter 5 helps students identify key stakeholders to whom they should direct messages of advocacy. Readers will learn how to use the power mapping tool to assess influential actors in a political landscape. Students will feel prepared to engage in authentic literacy practices as they direct messages of advocacy to influential stakeholders beyond school walls.

Chapter 6 shares SEL strategies for helping students navigate fear and uncertainty in the face of ineffective or absent climate policy. Readers will learn the value of using the CASEL framework to explore aspects of individual well-being and community well-being, as well as the role of writing in offering a "pressure valve" for difficult emotions. Students will feel prepared to engage in honest reflection about the emotional challenges associated with attempts to mitigate and adapt to new climate change realities.

Chapter 7 is devoted to project-based learning unit design that is organized around a meaningful driving question and connected to a real-world environmental problem. Readers will learn how to design unit "entry points" that encourage investigation, offer voice and choice in authentic products, and prepare students for collaborating during their investigation.

References

Bohlmeijer, E., Westerhof, G., Randall, R., Tromp, T., & Kenyon, G. (2011). Narrative foreclosure in later life: Preliminary considerations for a new sensitizing concept. *Journal of Aging Studies, 25*(5), 364–370. https://doi.org/10.1016/j.jaging.2011.01.003

Brundtland, G. H. (1987). *Our common future: Report of the world commission on environment and development.* Geneva, UN-Document A/42/427. https://sustainabledevelopment.un.org/content/documents/5987our-common-future.pdf

(2025, January 14). *Good and bad climate journalism with Denise Hruby*. The Salata Institute. https://salatainstitute.harvard.edu/good-and-bad-climate-journalism-with-denise-hruby/#:~:text=The%20journalist%2C%20who%20has%20worked,from%20political%20and%20industry%20leaders.&text=%E2%80%9CI%20don't%20see%20why,day%2C%E2%80%9D%20says%20Denise%20Hruby

Kimmerer, R. W. (2020). *Braiding sweetgrass: Indigenous wisdom, scientific knowledge and the teachings of plants*. Penguin UK.

Lake, O. O. (2024). *The story is in our bones: How worldviews and climate justice can remake a world in crisis*. New Society Publishers.

Nelson, R. G. (2022). How a virus exposed the myth of rugged individualism. *Scientific American*. https://www.scientificamerican.com/article/how-a-virus-exposed-the-myth-of-rugged-individualism/

Nixon, R. (2011). *Slow violence and the environmentalism of the poor*. Harvard University Press.

Smith, W., & Lewis, M. (2022). *Both/and thinking: Embracing creative tensions to solve your toughest problems*. Harvard Business Review Press.

Stibbe, A. (2021). *Ecolinguistics: Language, ecology and the stories we live by* (2nd ed.). Routledge.

Van Horn, G., Kimmerer, R., & Hausdoerffer, J. (2021). *Kinship: Belonging in a world of relations. Vol. 1: Planet*. Center for Nature and Humans Press.

Wang, T. (2016). Why big data needs thick data. *Medium*. https://medium.com/ethnography-matters/why-big-data-needs-thick-data-b4b3e75e3d7

1

Climate Stewards as Communicators

The first time I asked students to make a video documentary, we had been reading Tommy Orange's novel *There There*. Weaving together the stories of 12 characters from Native communities traveling to attend the Big Oakland Powwow, the novel upends stereotypical images of Native Americans by depicting them in a less familiar context. One of these characters, Dene Oxendene, receives a cultural arts grant to make a film that investigates what it means to be Native in urban cities, and to explore Native representation apart from the reservation setting many have come to expect to see in film, story, and history textbook depictions of Native Americans. As my students planned their films, we zoomed in on Dene's journey, reflecting on what it meant for him to create a storytelling opportunity for voices traditionally unheard. Thinking about how Tommy Orange's characters confront assumptions about the meaning of home and their own relationships with indigeneity helped my students interrogate the stereotypes associated with the places they described in their own films. Through reading *There There*, they thought more critically about what it means to reshape the terrain of what people imagine their home to be.

I have long thought of storytelling as a type of doorway, allowing us to access each other's perspective. Taking the comparison further, we can think about filming techniques as a door hinge, permitting a limited angle of rotation between two people who briefly have the opportunity to connect. While the viewing experience is ephemeral, the impact can be transformative—visual storytelling can help us convey the stakes of a story with stark concreteness. Making a film documentary challenges students to include imagery commensurate with the emotional significance of the topic, inviting decision-making about camera shots, voice-over narration, and editing. This chapter offers ideas for positioning students as climate stewards who communicate with effective messaging and draw upon powerful mentor texts for doing so. The suggested activities and included tools are structured to culminate in a video documentary project, but these resources can be pulled out and mixed and matched with other lesson ideas found in this book. To help students communicate with nuanced and sensitive advocacy, we must foster an appreciation for and an embrace of personal assets, caregiver assets, and community assets. To write with place-based awareness is not just knowing what is located in a place. It also means being open to what can be discovered in a place already mapped in your mental terrain.

When students feel daunted, remind them: no one else is a better documentarian of the place they call home. Years ago, a group of my students set out to inform about a methane gas power plant leak in their community. Their film began by directing viewers to look at an establishing shot of the sky, appearing beautifully lit and clear. The use of voice-over narration has the effect of a record scratch: there's nothing about the placid appearance of the Sun Valley skyline that suggests everyone living in this community is breathing the same toxic air. The next image displays power lines in the sky, directing the viewer's eye to the power plant located in the background. As the narrator describes how the health hazard had been covered up for years, he stresses the impossibility of protecting your loved ones from health dangers if you do not know that they exist.

Observing the very different places students wish to highlight reminds us that there is no single, monolithic way to tell the story of their community. Though some of these places are mere miles apart from each other, my students' individual storytelling intentions rendered them starkly unique.

To help students notice how an environmental story is being told, I introduce them to Kenn Adams' "The Story Spine." It's a popular, engaging improvisational exercise that can be done alone or in groups.

> Once upon a time.
> Every day . . .
> But one day . . .
> Because of that . . .
> Because of that . . .
> Because of that . . .
> Until finally . . .
> And ever since then . . .
>
> (p. 46)

In the hands of improvisational actors, the story spine can produce dazzling feats of mental invention. When actors are forced to build a narrative around materials assembled through whimsy and happenstance, the linear cause-and-effect structure offers simplicity and clarity. For complex narratives, though, the spine is too tidy. For our purpose, its simplicity is the point. By plotting familiar environmental problems using the story spine, we can then invite students to "fill in the gaps" and tell us what has been left out. I will model the story spine with an example based on how many people communicated about the wildfire crisis in Los Angeles, where I live, in January 2025.

Once upon a time, Southern California experienced a very dry winter.

Every day, people hoped it would rain, as it had not rained since the previous April.

But one day, the powerful Santa Ana winds fanned the flames of wildfires in different parts of Los Angeles, resulting in decimated neighborhoods and fatalities.

Because of that, first responders were challenged by empty water reservoirs and fire embers that traveled far and wide.

Because of that, more and more people had to evacuate their homes as more neighborhoods were threatened by the wildfire spread.

Because of that, residents in surrounding areas suffered harmful air quality.

Until finally, people pointed fingers at city officials, looking to blame someone.

And ever since then, residents have been on edge, frequently consulting their downloaded apps for wildfire maps.

For my students who experienced this story firsthand, they would recognize familiar story beats, fanned by reposted social media stories and fueled by fear, frustration, and anxiety. If we weave the same narrative with the counterstory stances, the story becomes less tidy, less linear:

In their book *Verified*, Mike Caulfied and Sam Wineburg offer suggestions for what to do before reasoning about a specific piece of content that reaches us through the web. These actions are summed up by the acronym: SIFT.

> Stop.
> Investigate the source.
> Find other coverage.
> Trace the claim, quote, or media to the original context.
> (2023, p. 13)

Students benefit from the SIFT method because they are pressed to distinguish between sharing sources (say, the person passing along the information to family and friends) and reporting sources (the news outlet where the person sharing found the item). I saw this in real time as they sought out new information about fire evacuation notices pertaining to their own neighborhoods or whether a loved one was directly in harm's way. When emotions run high, it is easy to pore over phone notifications obsessively and be more susceptible to clickbait headlines.

Using SIFT offered my students a type of fact-checking that is easy to learn and blunts the speed of an unfolding wildfire emergency narrative that assigns blame too quickly. A narrative had spread, blaming the inability of first responders to respond quickly on "budget cuts" made by the Los Angeles mayor. This narrative suggested that many residents are unaware about the responsibility the Los Angeles County Board of Supervisors also bears for disaster preparedness and response.

TABLE 1.1 Weaving Story Spine Details and Counterstory Stances

Story Spine Detail	Counterstory Stance	Fact-Checking
"Once upon a time, Southern California experienced a very dry winter."	Root Causes: What is the underlying cause of this problem?	According to UCLA researchers, "climate change may be linked roughly to a quarter of the extreme fuel moisture deficit when the fires began" (Madakumbura et al., 2025). "Climate Change A Factor In Unprecedented LA Fires" *Sustainable LA Grand Challenge*
"Because of that, first responders were challenged by empty water reservoirs and fire embers that traveled far and wide."	Systemic Responsiveness: How can we better anticipate this challenge in the future?	According to LADWP, the tanks' water supply needed to be replenished in order to provide enough pressure for the water to flow through fire hydrants uphill (Tidmarsh, 2025). Kevin Tidmarsh, "Fact check: What really happened with the Pacific Palisades hydrants?" *LAist*
"Until finally, people pointed fingers at city officials, looking to blame someone."	Beyond Either/Or: Why is it easier to assign blame to one person than to think about multiple factors challenging the disaster response?	"The supervisors are helped in part by the structure of their government, which unlike the city holds no single mayor-like elected official accountable for the entire body's performance" (Greene, 2025). "Why is Karen Bass getting so much blame for the LA fires but county supervisors so little?" Robert Greene, *Desert Sun*
"And ever since then, residents have been on edge, frequently consulting their downloaded apps for wildfire maps."	Collective Mobilization: How can we ensure we have access to timely, accurate emergency updates?	"At a time when misinformation is rife and public trust is low, many residents are ignoring official alerts and turning to the Watch Duty app for up-to-date information" (Jarvie et al., 2025). "L.A. residents get more erroneous fire evacuation alerts" *Los Angeles Times*

Once students understand there are choices made in what is emphasized in a climate story, we help them understand how storytellers offer a vantage point from which to view an issue. A story topic can be viewed in a panoramic, sweeping manner. It can be seen in up-close, granular detail. When I think about conversations involving fears about fire, I wonder what Traditional Ecological Knowledge (TEK) can teach us about fire-based land management practices and the role fire ecology plays in responsible land stewardship. Seeing online conversations about opportunistic landlords raising apartment rent near burnt-out areas led me to think about the emergency in terms of intergenerational awareness: how do we advocate for rent control, not just now, but in the future? When I saw Mutual Aid LA Network, a connector hub for mutual aid efforts offering ideas for volunteering and for receiving resources, it highlighted the importance of learning from community-guided solutions about how to help each other in our times of need. As we see in the case of the Los Angeles wildfires, narratives can spread quickly. We can help our students reject reductive ones that do not do the topic justice.

Video Documentary Planning Tool: From Topic to Call to Action

> **Box 1.1: Documentary Planning**
>
> Study Mentor Texts
> Develop Line of Inquiry
> Make a Community Asset Map
> Create and Circulate Surveys
> Conduct Interviews
> Make Hexagonal Thinking Connections
> Engage in SWOT Analysis
> Document Along the Way

Mentor Texts

The best resource I have encountered in helping students communicate as climate stewards is to expose them to powerful mentor texts. I remember the first time I read Allison Marchetti and Rebekah O'Dell's *Writing with Mentors* (2015)—the pages were swimming in Post-it notes full of my scribbled noticings. Each chapter beckoned me further, inviting me to see the "craft moves" that writers make. Mentor texts are writing models we share with students to show what strong writing looks like, so they can imitate style, language, structure in their own writing. Mentor text study offers an entry point to understanding the stakes of a story rooted in a place. A striking feature of communication devoted to climate stewardship is the frequent use of concrete, place-based details that direct attention to what is at risk, what is being threatened, what is slipping away.

If we want to help students navigate the media ecosystem with a clear-eyed understanding of the varied, unceasing bids for their attention, we need to help them evaluate content creation through identifying and understanding the craft moves content creators make. If we want to help students intervene and advocate for that which they love dearly, we need to help them to not only set out to combat misinformation but preemptively anticipate it. Modeling how we identify mentor texts for communicating as climate stewards empowers students to locate those "just right" texts on their own and fosters adeptness at naming the craft moves model communicators are using.

A critical feature of place-based calls to action is communicating how the story of a problem tends to be framed before communicating this *is* the story we are creating *now*.

Questions to Ask of Potential Mentor Texts for Climate Stewardship

Does it challenge an anthropocentric view of the world?
Does it call into question status quo practices that contribute to environmental harm?
Does it shed light on what is uniquely located in a place?
Does it promote system commitments on behalf of communities?

Does it provide evidence that draws upon multiple data sources?
Does it invite people to think and act like a future ancestor?

While selecting potential mentor texts, Marchetti and O'Dell ask themselves, "Does the text pass the highlighter text?" (p. 25). I love this question because it gets to the heart of why we select a mentor text—is there something in this text that is so well-crafted that we can find ourselves wanting to try out that move? Positioning students as curators of well-crafted texts is one way we can make the classroom more student-centered.

Finding Mentor Texts for Climate Stewardship

> **Box 1.2: Sources for Climate Stewardship Mentor Texts**
>
> Take a moment to consider the sources where you are likely to find mentor texts about climate stewardship. Jot down your list and display it for students so they can see how a collection is assembled from both likely and unlikely places! My list looks like this:
>
> Sammy Roth's *Boiling Point* newsletter
> Ian James, climate journalist
> *Grist* Magazine
> Katharine Hayhoe, climate scientist
> *Orion* Magazine
> Bill McKibben's *The Crucial Years* Substack newsletter
> José González, founder of Latino Outdoors
> Leah Green, author of *The Intersectional Environmentalist*
> *A Matter of Degrees* Podcast
> *Climate Action Now* Substack newsletter
> *Drilled* Podcast
> Isaias Hernandez, environmental justice educator and activist
> Redford Center film website

Being transparent about where you find your go-to mentor texts can help young learners gain an expanded sense of the information landscape. Just as I have found myself exploring previously unknown territory on various social media apps, students can find themselves visiting library databases and consuming climate news from legacy media sources typically hidden behind a subscription wall. Sharing my appreciation for these sources might broaden their notion of where interesting, smart, heartfelt writing and content creation exists.

One of the reasons I am so focused on identifying mentor texts that communicate effective place-based calls to action is we need to communicate the stakes of the story to our governing politicians who represent us. When politicians decide which issues to act upon, one of the deciding questions is this: "Is the position I am taking on this issue enabling me to be a megaphone for my constituents?" The better we become at communicating how climate stewardship empowers our local leaders to represent our interests, the easier it becomes for them to make clear to their community why effective governance means leading with intergenerational awareness and strengthening system commitments to our long-term flourishing.

Mentor Texts for Place-based Storytelling

The four mentor texts I describe in this chapter are some of my go-to mentor texts. They invite students to think deeply about how a call to action can weave personal narrative and place-based writing.

Essay Mentor Text: Emily Raboteau's "Spark Bird" (Orion Magazine)

Emily Raboteau's 2021 essay, "Spark Bird," offers place-based details (of location, weather, light, smell, sound, and species) that are evocative and memorable. I share this essay with students because Raboteau helps them notice subtleties about where they live and breathe and go to school. Conveying a call to action inspired by something observed locally, Raboteau guides her readers into awareness by bearing witness to multiple, simultaneous acts of disappearing.

Raboteau draws attention to the meaning of the phrase "spark bird"—that first bird that captures your interest and draws you into birding—only to gesture to her own unconventional "spark bird," a pair of owls painted on a storefront in Harlem. The mural is part of the Audubon Mural Project, a collaboration between the gallerist Avi Gitler, local property and business owners, and the National Audubon Society. After noticing it, she begins to notice all the painted birds along her two-mile walk to work at City College, mostly painted on the rolled-down gates of small shops on Broadway. The project aims to depict 389 birds: the number of North American species, according to the Audubon's 2019 birds and climate report, at risk of extinction from climate change. During her walks, Raboteau attempts to photograph them all.

As Raboteau considers what it means to bear witness to New York's endangered species, she offers facts about a particular bird species. Then she pivots to describing her own habitat. Braiding these stories, all located in one place, shows us how a street corner can be a type of palimpsest: a record registering appearances, vestiges, and disappearances. Her advocacy looks like a walk in the neighborhood with a camera. However, she guides us into awareness of what is at risk by tracking her noticings and making connections between natural and social phenomena occurring in the same place.

While reading "Spark Bird," we discussed how some parts achieved an effect that reminded us of the technique of defamiliarization: familiar objects are no longer perceived as such. A neighborhood walk suddenly offered transporting beauty in the form of painted murals. My students and I agreed that the place-based details that stood out the most were Raboteau's descriptions of the bird murals, particularly descriptors related to their size and color (which were often followed by a photograph). The idea that a mural might outlast a local shop demonstrates how powerfully art can register absence, even as it is felt sorrowfully. This image reinforces other images of loss, related to gentrification and pandemic closures, so powerfully woven throughout the essay.

Newspaper Editorial Mentor Text: Elizabeth Rush's "I Would Have Never Bought This Home if I Knew It Flooded" (New York Times, April 12, 2022)

TABLE 1.2 Mentor Text Move

Mentor Text Move:	Quotation:	How We Could Try It Out Ourselves:
Defamiliarize the Familiar	"At shops that have closed and not yet reopened, like the beauty salon with the laughing gull, the bird is always there."	Use defamiliarization to highlight what is at risk locally in our community.

In 2018, Elizabeth Rush published a Pulitzer Finalist book called *Rising* that documents how rising seas are transforming coastlines by weaving testimonials from members of vulnerable communities deciding whether to stay in place or relocate. Her experience makes her an apt choice for shedding light on the limits of protection offered by the Federal Emergency Management Agency in this powerful editorial. Rush reports that when FEMA asked the public how the government can do a better job of identifying flood risks and protecting homes and businesses, hundreds of people shared stories of how they were endangered by the agency's antiquated criteria and data.

When I share editorials with students, I draw attention to the use of the first-person voice. Climate stewards frequently write editorial articles in local newspapers. Editorials are a crucial means of sharing opinions about what is happening locally, about events or trends that are obscured by "noisier" national headlines. Because editorial writing gives students practice in expressing a particular viewpoint or stance on a topic, they serve as a great test run for other acts of advocacy. By keeping the unasked question, "Why does this matter?" in mind, editorial writers can keep frontline communities front and center, letting the audience know who is affected, who is most at risk.

Rush's reporting urges commonsense land use reforms, such as "mak[ing] disclosure of past flood damage and flood risk mandatory during all real estate transactions" and updating maps "to determine where it will flood today and in the future." By including the perspectives of those surveyed, she humanizes the data—revealing the heartbreak and regret that came with the realization that one's home was bought in a flood area where

TABLE 1.3 Mentor Text Move

Mentor Text Move:	Quotation:	How We Could Try It Out Ourselves:
Interrogating a Definition	"An affordable home shouldn't mean one that is cheaply built in an undesirable, unsafe location."	Challenge our audience to question what an abstraction means.

out-of-date flood maps do not account for floods exacerbated by climate change.

One reason I share this editorial with students is it brings to the fore the importance of surveys. Not only do we find out how those surveyed feel about a topic, we gain insight into their perception of the topic itself. The information gleaned from a survey can shed light on root causes perpetuating an environmental threat, as well as on barriers to participating in solutions.

Episode as Mentor Text: Isaias Hernandez's Teaching Climate Together Web Series: "Desert Ecology with Mojave Desert Land Trust" (2024)

Isaias Hernandez posts environmental education videos on social media under the moniker Queer Brown Vegan. One day, I was scrolling my Instagram feed when I saw him discussing desert ecology with members of the Mojave Desert Land Trust (MDLT). I followed the link to his YouTube channel, where I watched the full episode devoted to educating people about the connections between desert ecology, restoration efforts, and Indigenous knowledge. After a few minutes of watching, I realized I was watching a fantastic model for sharing information about a place beset by stereotypes.

When thinking about desert ecosystems, many people envision a barren wasteland. Hernandez lets various members of the MDLT dispel this notion by eliciting a layered, nuanced view of the Mojave Desert landscape, which is incredibly biodiverse. The MDLT maintains land acquisition and management protection of over 120,000 acres of California desert. From sharing images of the seed bank collection, where seeds are proactively gathered and stored for when they are needed to do restoration planting, to describing the unique soil surface, a web of cyanobacteria, lichen, moss, and fungus that enables this desert ecosystem to hold more carbon than upstate forest ecosystems, the MDLT team upturns the idea of a desert as dead and empty.

Some of my students become nervous at the thought of interviewing people when they do not feel "expert" enough to ask questions. This episode serves as a helpful mentor text for sustaining an interview—instead of dominating the conversations with his own questions, Hernadez makes room for the interviewee to answer fully, without interruption. When describing how we honor people with our listening, Nawal Qarooni emphasizes that an interviewer "must *inquire* before rushing in with their own commentary" (2023, p. 124). It can be tempting for the interviewer to pre-formulate a response, even while the speaker is in the act of sharing their story. Hernandez's conversations beautifully model how to let interviewees claim narrative space.

During the course of the episode, Hernandez interviews Kelly Herbinson, Executive Director of the MDLT, who sheds light on the conflicting goals of renewable energy planning and conservation efforts. The desert's abundant solar energy coupled with the perception that there is a lot of land with "nothing there" had made it the target of multiple solar development projects. Herbinson explains how the Desert Renewable Energy Conservation Plan, a landmark collaboration between multiple agency stakeholders, has outlined development focal areas where solar development is restricted to areas that have been identified as not as ecologically important as others. This interview segment offers insight about what it means to go beyond Either/Or thinking and balance environmental solutions that compete for the same space.

TABLE 1.4 Mentor Text Move

Mentor Text Move:	Quotation:	How We Could Try It Out Ourselves:
Offering an alternative to "siloed" problem-solving	At the 13 minute 24 second mark of the film, Hernandez says, "We're not trying to say we're anti-renewables."	Highlight the benefits of implementing environmental solutions simultaneously and in the same place.

Video Mentor Text: Faith E. Briggs's This Land (2020)

Since students usually have little experience planning a video documentary, *This Land* serves as a great introduction to framing personal stories with a purposeful use of camera angles and voice-over narration. Faith E. Briggs powerfully models how documentary storytelling can underscore the personal, exploring how outdoor spaces like parks and public lands can feel unwelcoming to historically marginalized communities. As images of her running on a track are juxtaposed with images of her running through Grand Staircase Escalante National Monument, we hear her explain in the voice-over narration that being a conservationist is an identity explored through running.

Watching Briggs navigate the ruts, inclines, and downhills as she runs through three national monuments—public lands protected under the Antiquities Act—makes for a vivid and immersive viewing experience. Her exploration of what it means to be a conservationist amid the threat of rolled-back protections for public lands is a layered story, one that prompts us to consider all the barriers to our own access to the outdoors. Whether through speaking directly to the camera or providing voice-over, Briggs invites us to see ourselves in America's public lands. As we watched scenes of her trail running with local activists through national monuments at risk of losing territory, my students and I tracked our noticings about this 11-minute documentary short.

As Briggs and her running companions pore over maps, a line is drawn on a map as they plan their 150-mile run, beginning at Cascade-Siskiyou in Oregon, snaking down through Grand Staircase-Escalante in Utah, and ending at Organ Mountain-Desert Peaks in New Mexico. A student pointed out that the concept of redrawn lines echoes throughout the film in visual and verbal form, as the then presidential administration attempted to redraw the boundaries for public lands. The camera footage shot overhead as Briggs runs often reveals roads and trails that parallel the lines of waterways and mountain ridges, depending on the camera's vantage point. Most strikingly, combining public land advocacy with her passion for running redraws the image of conservation.

TABLE 1.5 Mentor Text Move

Mentor Text Move:	Quotation:	How We Could Try It Out Ourselves:
First-person Narration	At the 30-second mark of the film, Briggs says "The battle now is saying, 'No, I am a conservationist and redefining what that means'."	Highlight the value of telling our own story in an unmediated manner.

Activity: First-person Voice-over

As Briggs describes being alone and feeling lost at Grand Staircase-Escalante, her voice-over narration begins with a mix of camera shots. To practice planning the types of camera shots you will use in your own film, write narration commentary for voice-over. Include setting description with at least one establishing shot: a first shot in a scene that is used to convey basic facts about the setting.

Activity: Interview About a Place Worth Preserving

When we encourage students to draw upon their personal assets, we exalt the cultural, linguistic, and familial resources they bring from outside the classroom. Highlighting conversations students can have with their caregivers is a way to affirm their family's and community's funds of knowledge (Flores & Fránquiz, 2023). To help students improve their communication about places in need of advocacy, enlist their caregivers as co-creators of stories that draw upon repositories of shared memories. In her book *Nourishing Caregiver Collaborations* (2023), Nawal Qarooni explores the power of oral conversations to grow ideas and connect us: "The first step of the writing process—talking about ideas aloud—is a clear entry point for caregivers of all cultures and abilities to engage" (p. 128). She suggests using inclusive language for framing family interview assignments so as not to alienate or exclude students with different family makeups: "*your grownups, community,* and *chosen family.*"

Directions: Interview a caregiver or your grown-ups about a place you both know. The questions offer an opportunity to compare and contrast your attachments to a place.

When you think of this place, what images come to mind? (What is the "movie" in your mind?)

Why is this place emotionally resonant for you? How does it offer you a sense of connection?

If you were to "map out" other places that are emotionally resonant for you, what would these places have in common?

Activity: Tree Rings Map

To help students prepare mentally for their interviews, I invite them to create a tree rings map. The mapping encouraged here is nonlinear, marked by personal symbols—I encourage students to think of places that hold emotional significance for them and place them within a set of tree rings. Tree rings, visible as concentric circles when the tree trunk is cross-sectioned, provide information about how old the tree is and what the weather was like during each year of the tree's life. I love this map activity because it honors the students' choices in what they wish to share about themselves. Just like rings of a tree trunk indicate growth response and sensitivity to varied climate conditions, student mapping of their own lives reveals their perception of those places, haunts, and refuges that made an impact.

Directions: Create a tree rings map depicting the radiating influence of places on your sense of identity and life.

My student Austin's tree rings map and accompanying narrative beautifully illustrate the gifts this type of place-based awareness can offer:

The same water that turns an egg hard makes a potato soft. Environment is everything. I am a product of my surroundings. And though these places hold space in heart, it is the people that make them special.

This tree ring, in particular, showcases the places I share with my favorite person—my dad. At the center is the day we spend the most time together, consistently: every Sunday, 9 AM, Almansor Park in Alhambra, California. Soccer practice.

Here, an eclectic mix of characters—bonded by ethnicity, gossip, and a love for the game, gathers. The core pillars of my identity—humor, sharp wit, and the love of spontaneous conversation take root in this field. This place has given me more than

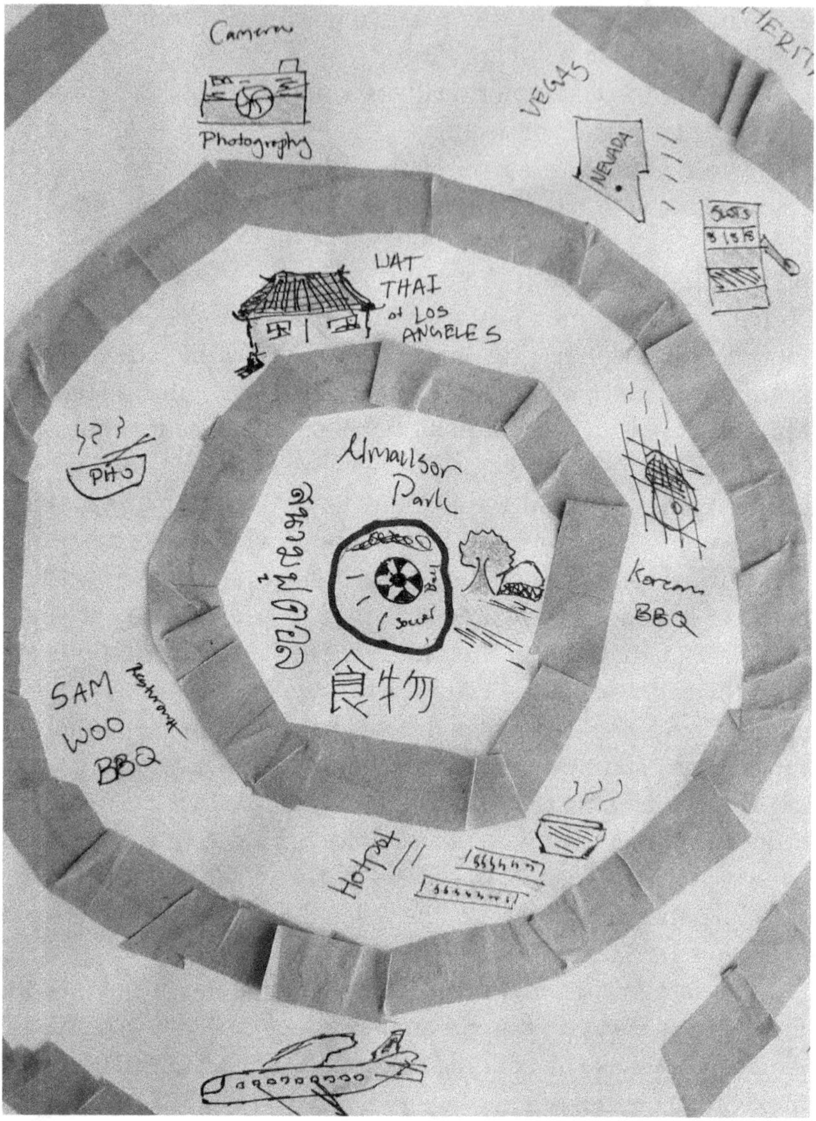

FIGURE 1.1 Austin's Tree Rings Map. Photograph by the author.

just athleticism and technique. It has taught me to hold my own in a room of loud voices, to cooperate for a cause, to connect with those who share little in common.

After soccer comes the meal—Chinese Diners like Sam Woo, phở, hotpot, or wherever our cravings lead. But the spirit of the pitch follows us. The conversations carry on. The second half of the game begins. And in a sense, the game never really ends. I move through the world the same way I move through these places—unafraid to speak, quick to joke, and seeking a good conversation.

Box 1.3: Filmmaking Resources

Some Filmmaking Contests and Resources to Share with Your Students
One Earth Film Festival—Young Filmmakers Contest
The Redford Center Youth Stories
National Geographic Slingshot Challenge

Launching the Video Documentary Project With a Guiding Question

Directions: You and your group will make a video documentary advocating for a solution based on an identified community need. Your focus will be guided by your curiosity, but you will engage in activities to help you narrow the scope of your inquiry. Keep these questions in mind:

> What community assets exist that people might not know about? What already exists that can be leveraged in a different way to address the need?
> What are some root causes of the problem? What is the difference between the problem's symptoms and root causes?
> What insights from people who are experiencing this issue or working on this issue are helpful to settling upon a solution?
> What were your original questions? How have some of your questions/thoughts shifted during this learning journey?

It is up to your group how you want to structure your video, but some visuals related to your data points need to be included.

Quantitative data: Consider making a graph, table, or timeline to convey quantitative data.

Qualitative data: For important information that cannot be captured numerically (examples: open-ended survey responses, descriptions of people's feelings based on interviews), take detailed notes and save audio recordings of interviewees.

Box 1.4: Video Tools

Some tools for video editing: InShot, CapCut, iMovie, Picsart, Videoleap, Canva

Some tools for creating data visualizations: Keynote, Excel, Google Sheets, Canva

Some tools for creating surveys: Google Forms, Mentimeter, Survey Monkey, Slido

---

Activity: From Idea to Question

Directions: As we begin to ponder the focus of our video documentary, you will document the journey of your learning: document moments of inquiry that you will track as your questions shift and evolve. The concept that will launch your journey is "change." Write down as many words that come to mind as your group ponders what needs to change in society:

> Think about the problems you identified and what needs to change (a pressing need in your community, a problem in society). Write the problem in the form of a question as you begin to narrow in on the root causes of the problem:

Example: How does my community protect habitats for endangered species?

Choose a non-classmate to survey. Inform them of one of the needs you identified and find out what they think about the problem.

What do you think are the root causes of this problem?

What assets exist in the community that could be used to address this problem?

Do you think there should be different short-term and long-term approaches to addressing this problem? What are some examples?

Community Asset Mapping is a tool that helps us identify the hidden wealth of knowledge inherent in our communities. Instead of looking at our communities through a deficit lens—noticing what's missing or broken—the asset-mapping tool helps us notice and appreciate the assets already existing in our communities.

How can the resources already existing in communities be directed to meet needs? Community Asset Mapping helps us find these assets and connect people with untapped resource providers. No longer focused merely on what communities lack, asset-mapping helps us imagine existing features as assets to be utilized differently.

Activity: Community Asset Reflection

What is a pressing community need?

What organizations, institutions, businesses, and individuals might be interested in partnering and providing support?

Thinking about the need you selected, spend some time researching relevant organizations and community groups. What organizational alliances could be built with other potential institutional allies?

Activity: Hexagonal Thinking

One way to help students weigh solutions is to use the hexagonal thinking strategy. While working as a team, it's hard to let ideas marinate before settling upon a project focus—inevitably, there is an impatient group member who thinks the solution is self-evident and requires no discussion. Designed by curriculum designer Betsy Potash, hexagonal thinking is an activity that helps students deepen their thinking by visually connecting a series of words on hexagon tiles around a theme.

Directions for students: Record terms, concepts, or ideas on hexagonal tiles and arrange them as you perceive relationships between the tiles, creating a visible web of connections. Using the hexagonal thinking strategy is a way to help us broaden the scope of our understanding. After gluing tiles to a paper, write questions that emerge based on discussing the tile relationships. (Teachers: those of you made anxious at the thought of cutting hundreds of hexagon shapes out of paper, fear not! Hexagonal tiles intended for sewing can be purchased in bulk easily.)

Students jump into this activity with gusto. Maybe because they know there's no "right question" to ask, they immerse themselves in possible mental combinations easily. The activity sets them up to work more autonomously as they move towards research investigation because they prioritize questions among the question set they generated.

Example: One group wrote "preventing overconsumption" on a center tile. They then wrote these terms related to their topic on separate tiles: supporting local farmers; farmers markets; regulating fast food intake; reducing food waste; and living sustainably. The questions spurred by juxtaposing these tiles were:

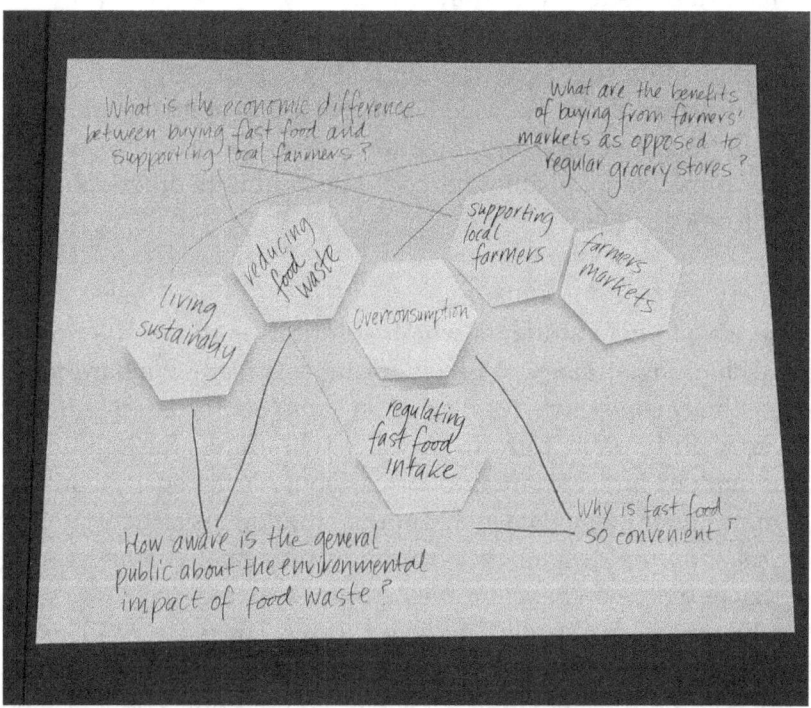

FIGURE 1.2 Hexagonal Thinking Map. Photograph by the author.

What are the benefits of buying from farmers markets as opposed to regular grocery stores?

What is the economic difference between buying fast food and supporting local farmers?

Why is fast food so convenient?

How aware is the general public about the environmental impact of food waste?

As they plan their video documentaries, I ask students to track how their questions are shifting and evolving as they become more knowledgeable about their topic. The hexagonal thinking activity is a useful way to leverage physical manipulatives to make connections and "bump" ideas against each other. This group's initial guiding question—how are we able to reduce food waste?—has branched out into new avenues of thought.

Survey Creation

Activity: Invite your student to design a survey that will help them weigh both quantitative and qualitative data. The survey should include both open-ended and close-ended questions. Open-ended survey questions allow respondents to give free-form answers without any restrictions. Close-ended survey questions produce quantifiable data (such as asking respondents to choose a number on a scale correlating to options such as "Deeply Concerned" to "Not Concerned at All").

When my students create their surveys, they push the survey out to at least 30 people. A student group dedicated to reducing Fashion Overconsumption developed the following survey questions:

How do you define "Fashion Overconsumption"?
To what extent do you think fashion overconsumption affects the environment?
How do you think fashion overconsumption affects the environment?
Who do you think is to blame?
How do you think we could solve this problem?

To encourage student groups to get the most out of reviewing survey responses together, use this conversation tool as a way to sift through the data:

> Which responses were most illuminating?
> Which responses were confusing?
> What were "the common denominators"?
> What is not captured in these survey responses?

Through identifying and articulating gaps between what the group understands about the identified problem and what respondents understand, new action steps can be decided upon. Together, students can determine trends in the responses, as well as outliers. Discerning a pattern in the responses can point the way to the best storytelling moves to use as they plan their video. Based on a survey response, my student Maddie devised a memorable analogy that made the final cut of her group's documentary:

> Think about it like this: in a colony of ants, if just one ant gets lost or leaves their normal line, it throws the whole system off. Similarly for us, if we could start changing our habits, then even the CEOs of their companies would have to start changing their business practices.

Activity: SWOT Analysis

When it comes time for group members to make decisions about the best solution for their identified community need, they may be torn between what seem like equally viable options. This is when SWOT Analysis can play a helpful role. As groups assess each possible solution, they can raise these questions: What are its strengths? What are its weaknesses? What opportunities exist with pursuing this solution? What threats exist to implementing the solution? SWOT was created to help organizations develop a full awareness of all the factors involved in making a decision. Students can weave SWOT with the storytelling stances to make their assessment more specific, as seen in the following examples.

Systemic Responsiveness: What oversight exists to ensure water permits are enforced? (Weaknesses)
Root Causes: Does fear of budget cuts inhibit our planning for an awareness campaign? (Threats)
Beyond Either/Or: Can solar panels placed in vineyards reduce water usage and provide grapes shade from intense heat? (Strengths)
Kinship and Reciprocity: Can learning more about mutualistic relationships between plants and insects help us prioritize our conservation efforts? (Opportunities)
Intergenerational Awareness: Can we start outreach about media literacy in elementary schools to reach students at a younger age? (Opportunities)
Collective Mobilization: Are there social media influences we can enlist to help spread awareness? (Opportunities)

Documentation

To meaningfully assist students with tracking their evolving ideas while developing their video project, I teach them the practice of pedagogical documentation. Angela Stockman explains the role of documentation in making learning visible in her book, *The Writing Teacher's Guide to Pedagogical Documentation: Rethinking How We Assess Learners and Learning* (2024). An educator and professional learning facilitator who has worked with students at every age, Stockman invites students to document their discoveries in a way that is textured and layered. This documentation is often multimodal in composition and composed of stories that allow teachers to bear witness to a journey of learning. The dedicated act of gathering artifacts—notes, photographs, videos, sketches, and audio recordings—assists students with their own story making.

In project-based learning, it can be challenging for learners to gauge their own growth—the journey of the project from point A to point B can be full of false starts and revisited drawing boards, and those moments can be frustrating. But those moments of frustration are also full of meaning. Usually, they mean our first attempt or theory was a bit undercooked, and we need to spend more time letting our ideas simmer.

FIGURE 1.3 My Student Julienne's Student Scrapbook Cover Illustration. Photograph by the Author.

At the beginning stage of video planning, I gave every group of students a wire-bound scrapbook and offered these instructions:

> You're assembling a scrapbook that is a testament to your thinking on this learning journey. A documentation scrapbook makes the process, not just the product, of learning visible. This is a way to lean into the stories students are telling. You will include artifacts from your learning—images, drawings, reflections—anything that captures your thinking and the shifts in your thinking.

One way to begin is to invite students to engage in image collection in their communities, mindful of the visual imagery that is captured by their lens. When comparing their gathered images, we noticed how certain objects appeared repeatedly in their photographs: murals, signage, plants, animals, transportation routes, architectural relics, new building development, patterns in the landscape, maps, graffiti, and green spaces. Many students include photographs in their scrapbooks, as markers of shifting in their intended story plan. Because of a new idea sparked by something they now want viewed in their film, they chose to plot their story in a new direction.

By observing what students chose to include in their scrapbooks as evidence of key moments in thinking about their video documentary topics, I learned which tools and activities were helpful to them. My student Kailo's scrapbook reflection inspired her group to think about fast fashion alternatives.

FIGURE 1.4 Scrapbook Reflection About Fashion Overconsumption. Photograph by the Author.

Kailo's Scrapbook Reflection

Fashion has always been something I'm passionate about and is mainly how I express myself. But it's easy to get carried away in all the media/micro trends, especially with how quickly everyone moves on from things. Because I love fashion so much, sometimes it's easy for me to overconsume. I want fashion to be something that makes me feel proud of myself instead of something that makes me feel guilty. To combat this, I wear clothes in different ways (a skirt as a shirt), thrift, wear my parents' clothes, support small businesses, and always try to find new ways to be able to express my feelings through fashion.

Having Clarity About Our Storytelling Motives

Marshall Ganz created the Public Narrative framework based on the stories told by social movement leaders. In his book *People, Power, Change* (2024), Ganz outlines a framework for public narratives that links the Story of Self, the Story of Us, and the Story of Now. A Story of Self communicates the values that have called you to leadership; a Story of Us communicates the values shared by those in action; a Story of Now communicates an urgent challenge to those values that demand action now. In contrast to the Story Spine structure that sequences story elements in terms of linear cause and effect, the Public Narrative framework offers an audience a choice and an opportunity to act on those values that link all three stories.

This storytelling framework offers a compass to student storytellers. Having investigated, interviewed, surveyed, and fact-checked, students can feel overwhelmed by the volume of information they have gathered. By giving them an opportunity to develop their Public Narrative, they gain a scaffold for articulating how their message of advocacy connects with their values.

Activity: After students have spent time learning about the community need they identified, share Ganz's framework to help them make sense of their data and connect their recommended actions with their emotions.

Story of Self: What are particular moments in your life story from which your motivating values arise?
My student Alondra writes: "I instantly thought of how the U.S. participates in overconsumption. I noticed the difference between the waste my family produces in the U.S. versus when we are in Mexico. In Mexico, we tend to mainly purchase basic ingredients and those that are in season, which is partly due to the limited options at a local store."

Story of Us: Does the story you are narrating make you feel more connected?
My student Charlie writes: "Yes, I've never had a formal way to express the criticality of media literacy, especially in a way that doesn't discredit the knowledge people do have, and this gives me the opportunity. I've always felt entrenched in the digital world as it pertains to social media and other mediums of content that arise, so being given a voice in what feels like a neverending echo chamber makes me feel heard."

Story of Now: Do we feel compelled to act because something we hold dear is threatened? Do we know what challenge we're asking people to act on?
My student Yijing writes: "By asking others to educate themselves on the crisis, we are asking people to change their way of life."

Conclusion

These responses bear witness to the fact that developing communication skills as a climate steward means learning how to tell a story on *behalf* of someone else. Witnessing students become more attuned to their motives for choosing their story topics helped me better understand how important it is that they have the opportunity to be the generator of stories, not merely recipients of the ones I have chosen. Devoting time to practicing fact-checking and data-gathering enhances students' abilities to convey the stakes of their stories in a manner that does them justice.

> **Box 1.5: Educator Spotlight**
>
> **A Peek Into Their Practice**
>
> "That's what is key to this journey. Honestly caring about what kids are bringing to the table."
>
> Heather Zajdel is a secondary science teacher who has taught in the Philadelphia school district for 17 years. Her recent teaching experiences have been in nontraditional, alternative high school settings, where she has become acutely aware of how factors outside of class can weigh on students and make school less of a priority. Noting the high degree of student transience, she admits, "There is a pressure to make lessons extra meaningful because of all the things weighing on them."
>
> As a result, Heather honors the stories students want to share about themselves, stories about being nourished by the familiar. Making space for rhapsodies about the most perfect salty, tangy, crunchy exterior can create passionate debate about the ideal snack. One of her favorite collaborations has been teaming up with a Japanese teacher, Masaharu Shigematsu, and setting up a Padlet board where students from both their classrooms could post pictures of food they were eating, circumventing the challenges of differing time zones. Heather explains: "It became this show-offy moment of here's something I really care about and want you to see. That was the opener: I know something, and I want to share it. That built some curiosity. It wasn't 'Here's a problem we want to fix.' The opener was 'Here's something I care about'."
>
> This cross-cultural show-and-tell lays the groundwork for her students' inquiry, beginning with the question "What is it that I like?" which inspires more questions: "Where does it come from? Is it grown here? How did it get to my plate?"

In one instance, talking about burgers led to talking about cows, which led to talking about methane. Suddenly, storytelling about how their favorite food made it to their plate led to a wide-ranging conversation about the fragility of the environment that was propelled by speculative musing aloud. To encourage student storytelling about food footprints and their impacts, consider using Heather's prompt to frame the conversation.

"Instead of leading with the carbon cycle, we begin with talking about our favorite food, and why we love it."

Heather's prompt:

Talk about a meal from home that's important to you.

Take apart the components of it. Where did it come from? What was the carbon footprint of it? What did it mean to the environment?

Try and think of a version of it that would still be culturally meaningful to you and your family but might reduce the toll on the environment. Make a version of it that is more environmentally friendly.

Guiding Question: What is an environmentally friendly version of your favorite meal?

Learning Standards:

Next Generation Science Standards

HS-ESS3–3—Earth and Human Activity

Evaluate or refine a technological solution that reduces impacts of human activities on natural systems.

HS-LS2–5 Ecosystems: Interactions, Energy, and Dynamics

Develop a model to illustrate the role of photosynthesis and cellular respiration in the cycling of carbon among the biosphere, atmosphere, hydrosphere, and geosphere.

HS-LS2–7—Ecosystems and Interactions

Design, evaluate, and refine a solution for reducing the impacts of human activities on the environment and biodiversity.

ELA/Literacy Standards

CCSS.ELA-Literacy.SL.9–10.1.c

Propel conversations by posing and responding to questions that relate the current discussion to broader themes or larger ideas; actively incorporate others into the discussion; and clarify, verify, or challenge ideas and conclusions.

CCSS.ELA-Literacy.SL.9–10.4

Present information, findings, and supporting evidence clearly, concisely, and logically such that listeners can follow the line of reasoning and the organization, development, substance, and style are appropriate to purpose, audience, and task.

Connections to Crosscutting Concepts:

Structure and Function—Relationship between crop varieties and different growth conditions

Cause and Effect—Effect of canned and frozen food (preserving processes) on appearance, quality, and flavor

Systems and System Models—Impact of changes in one part of the food production system on the whole

References

Adams, K. (2010). *How to improvise a full-length play*. Simon and Schuster.

Caulfield, M., & Wineburg, S. (2023). *Verified*. University of Chicago Press.

Flores, T. T., & Fránquiz, M. E. (2023). *Cultivating young multilingual writers: Nurturing voices and stories in and beyond the classroom walls*. National Council of Teachers of English (NCTE).

Ganz, M. (2024). *People, power, change*. Oxford University Press.

Greene, R. (2025, January 25). Why is Karen Bass getting so much blame for the la fires but county supervisors so little? *Desert Sun.* https://www.desertsun.com/story/opinion/2025/01/25/unlike-bass-la-county-supervisors-take-little-heat-for-the-fires/77889787007/

Hernandez, I. (2024, October 23). *Teaching climate together: Desert ecology with Mojave desert land trust* [Video]. YouTube. https://www.youtube.com/watch?v=eeZ_HXtXbEY&t=80s

Jarvie, J., Toohey, G., & Castleman, T. (2025, January 10). L.A. residents get more erroneous fire evacuation alerts. *Los Angeles Times.* https://www.latimes.com/california/story/2025-01-09/emergency-alert-text-message-los-angeles-fire

Jolly, C., & Hasset, W. (2020). *This land* [Film]. Mellifera Collective. Vimeo. https://vimeo.com/394088858

Madakumbura, G., Thackeray, C., Hall, A., Williams, P., Norris, J., & Sukhdeo, R. (2025, January 13). Climate change a factor in unprecedented LA fires. *Sustainable LA Grand Challenge.* https://sustainablela.ucla.edu/2025lawildfires

Marchetti, A., & O'Dell, R. (2015). *Writing with mentors: How to reach every writer in the room using current, engaging mentor texts.* Heinemann.

Orange, T. (2018). *There there.* Alfred A. Knopf.

Qarooni, N. (2023). *Nourishing caregiver collaborations: Elevating home experiences and classroom practices for collective care.* Stenhouse.

Raboteau, E. (2021, February 19). Spark bird. *Orion.* https://orionmagazine.org/article/spark-bird/

Rush, E. (2022, April 11). Opinion | As climate changes, flood rules need reform. *The New York Times.* https://www.nytimes.com/2022/04/11/opinion/climate-change-flooding.html

Stockman, A. (2024). *The writing teacher's guide to pedagogical documentation: Rethinking how we assess learners and learning.* Taylor & Francis.

Tidmarsh, K. (2025, January 9). Fact check: What really happened with the Pacific Palisades water hydrants? *LAist.* https://laist.com/news/climate-environment/why-did-pacific-palisades-water-hydrants-run-dry

2

Climate Stewards as Systems Thinkers

Early in my teaching career, I asked my students to create a PSA on an environmental topic. After reflecting on how it went for them, I asked myself: Why didn't I ask them to create a *real* PSA? Sometimes, the student work done in our classrooms is not as relevant or cognitively demanding as it could be due to the artificiality of the culminating unit tasks. We don't always take advantage of the opportunity to create multiple audiences for their learning and empower them to deliver message campaigns beyond our classroom walls. Unquestionably, there is huge value in simulating and practicing real-world communication, but role playing should not be both the means and the end when it comes to climate stewardship. To create the multiplying effect needed to mitigate the worst effects of our climate crisis, we need to stage conditions that allow students to decide what is the best intervention to make, just as policymakers would. Educators can design lessons that enable this learning progression—the shift from merely playing the role of problem solver to really owning the identity of problem solver through words and actions—by fostering authentic literacy practices. Connecting students with real audiences for their communication is critically important given the need for building climate stewardship capacity. This chapter outlines a potential unit of study that will help educators improve their lesson design for problem-solving: specifically, how to prepare learners to broaden their view of a problem with the help of others. The suggested activities culminate in the creation of a podcast episode that serves to share knowledge with an audience about potential solutions. This podcast conversation offers a metacognitive opportunity for learners to account for personal shifts in their own thinking about these problems.

Why Systems Thinking?

Often, educators seeking to make environmental problem-solving a mainstay of their classroom field some unexpected curveballs. Despite our best efforts, the solutions pitched by our students might be hastily articulated or superficially conceived. Owing to a problem's chronic occurrence or complexity, it can be difficult for problem solvers of any age to come up with a result that goes beyond a quick fix. Dismayed by what we may perceive as our own lack of expertise, educators and students alike might shy away from discussing environmental problems, rationalizing there are better minds tackling these issues elsewhere.

Bringing systems thinking into the classroom can both build student confidence in approaching problems and build educator resourcefulness in helping learners navigate a landscape of possible solutions. Tackling the challenges associated with mitigating and adapting to climate change will require us to move away from trying to address just one part of a system. Introducing systems thinking to students builds a habit of assessing a variety of environmental problems through the concept of reinforcing and balancing loops, which critically illuminates how parts of a system interact and affect each other over time. Helping students understand that single events do not occur in a vacuum but rather occur in a complex interplay of events prepares them for understanding the effects of climate change already underway and how to implement effective mitigation responses.

In contrast to traditional mechanistic thinking, a systems thinking approach emphasizes a holistic understanding of a problem by thinking about how to optimize the performance of an entire system (Meadows, 2008; Stroh, 2015). This approach assists learners in appreciating the vast number of causal relationships that define our world. It supports a fundamental practice of identifying the interlinking elements that contribute to the challenges we face (Wiel, 2016/2021). For these reasons, systems thinking competency is an invaluable aspect of developing climate change literacy. Not only does it foster "the ability to collectively analyze complex systems across different domains (society, environment, economy, etc.) and across different scales (local to global)," but it lays the groundwork for problem-solving that is integrative: not focused on single parts of a problem (Wiek et al., 2011, p. 207). For learners overwhelmed by the enormity of the climate crisis, it can be difficult figuring out where to start. The concept of system feedback offers a helpful entry point for developing a big picture way of looking at connections in the world to better decision-making. To help educators demystify the practice of systems thinking, this chapter offers a guide for cultivating these skills:

- The ability to understand complex dynamics of a situation or problem
- The ability to identify high-leverage interventions
- The ability to reflect on one's own role in exacerbating the problem
- The ability to share knowledge with an audience about actionable solutions

System Feedback Loops

To begin, it is helpful to expand the way most students think of feedback: as a response to a particular performance or activity so improvement can be made. System feedback is information that returns to its original sources such that it influences that initial source's subsequent actions. Feedback loops are circular processes that amplify or dampen the effects of an action or change. These loops can be reinforcing or balancing, depending on whether they increase or decrease the gap between a desired state and an actual state. Let us look at some feedback loop (also

Climate Stewards as Systems Thinkers ♦ 49

known as a causal loop) diagram examples to better understand the circular stories they can represent.

The Reinforcing Causal Loop

Change occurring with one variable in the system sets the stage for change throughout the loop. This loop is reinforcing because the increase seen with one variable ("increased coral bleaching") continues in the chain of causality. So, "reinforcing" means an increase in change in the same direction:

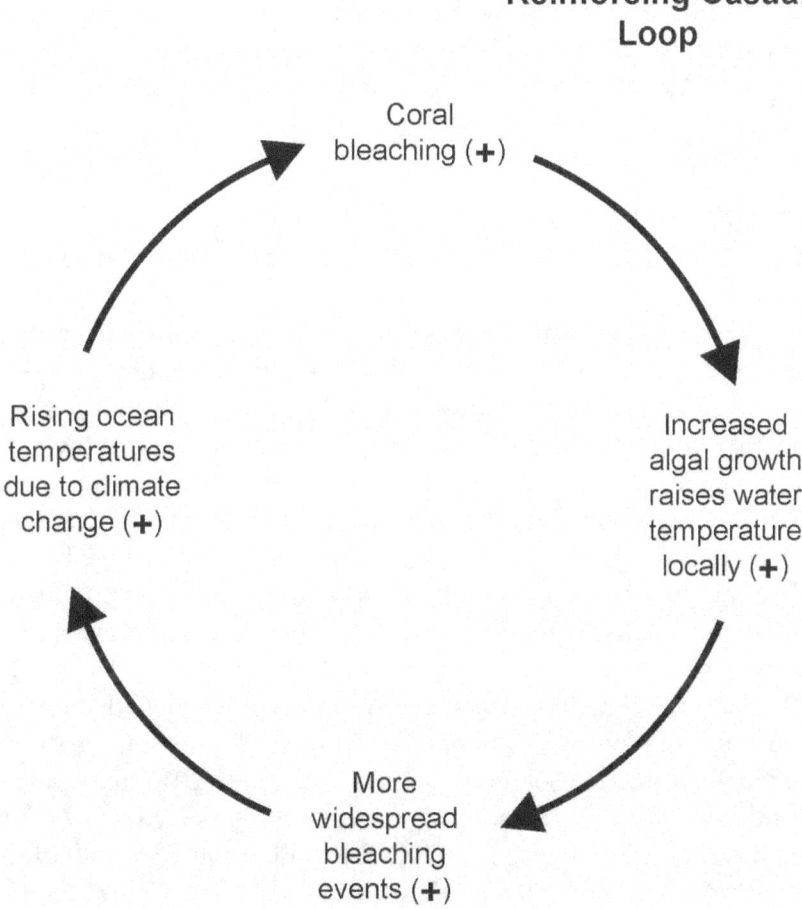

FIGURE 2.1 Reinforcing Causal Loop Diagram.

A reinforcing loop is a story of amplification—change with one variable ("rising ocean temperatures due to climate change") sets off a chain of increases that forms a circular story, ultimately offering reinforcing feedback to the source element ("more widespread bleaching events"). When we present examples of causal loops to students, it is helpful to engage in "think alouds" as we explain what happens as a result of one changing variable in a loop. Educators should stress to learners that understanding circular causal relationships will increase understanding of complex systems, because we are not just fixated on one story element. Instead, we are taking the long view and thinking about how elements within a system impact each other. The visualization tools associated with systems thinking allow us to better appreciate the interdependent nature of our world.

The Balancing Causal Loop

The causal loop diagram depicted demonstrates a balancing feedback loop.

As more greenhouse gas emissions are poured into the atmosphere, more heat is added to the global system, increasing negative impacts—extreme weather events, sea level rise, health emergencies, poverty. The (+) sign indicates how the second variable changes in the same direction as the first. As a result of these dangerous impacts, worry grows, leading to actions to shrink emissions. These actions, materializing through legislations, treaties, and personal behavior, lead to fewer emissions. The (−) sign indicates how the second variable moves in the opposite direction as the first, closing the feedback loop. Balancing loops tend to counteract growth in systems; in other words, as emissions come back down, it creates a balancing effect. Researcher David Peter Stroh (2015) makes the difference between these types of causal loops clear: "In contrast with reinforcing feedback loops, which *amplify* an existing condition, balancing feedback seeks to *correct* or reverse a current state by bridging a gap between actual and desired performance" (p. 50).

Climate Stewards as Systems Thinkers ◆ 51

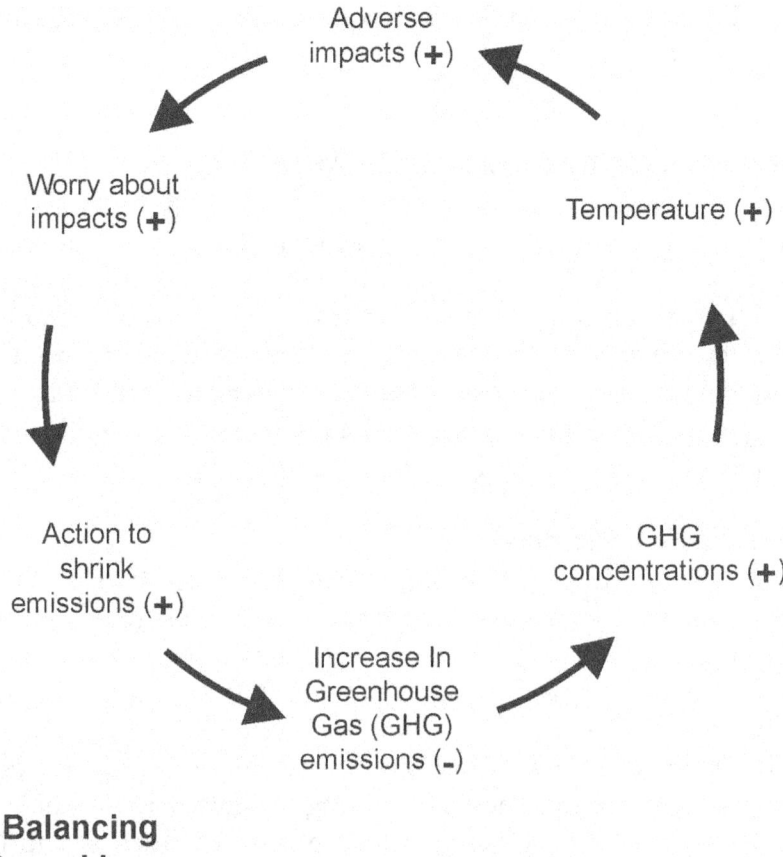

Balancing Causal Loop

FIGURE 2.2 Balancing Causal Loop Diagram.

Activity: Reflect on the Influence of Information Feedback

As teachers aim to help students become climate stewards, they should incorporate causal loop diagramming into the exploration of how climate change emergencies play out through systems (both natural and social). After introducing the concepts of reinforcing and balancing feedback loops, reflect with students about the influence of information feedback.

> How do feedback loops help us understand the implications of our behavior?

What information feedback would cause us to change our behavior?

Activity: Create a Causal Loop Diagram

Practice making your own causal loop diagram representing the impact of a keystone species within an ecosystem. To create such a diagram, begin drawing an arrow from the cause and then point the arrowhead to the effect. Use the following organizer to help you explore the relationship between related variables. On the right side of the chart, describe that variable's role in maintaining a balanced species population.

Finding High-leverage Points

In systems thinking, a leverage point is a place in a system's structure where a solution element can be applied (thwink.org). This place is a low-leverage point if a small amount of change force causes a small change in system behavior. It is a high-leverage point if a small amount of change force causes a large change in system behavior. For example, offering students a college tax credit would be a low-leverage point whereas meaningfully addressing the rising cost of college education would be a high-leverage point—affecting structural change in a system. As the seesaw image in Figure 2.3 attests, finding the highest leverage means the systems thinker is not settling for superficial "Band-Aid" fixes:

An important aspect of helping learners become responsible climate stewards is articulating the difference between the root causes of our climate crisis and the symptoms that show up as signs of these root causes. Guiding them into the practice

TABLE 2.1 Planning a Diagram

System Variables:	Effect of Variable Within an Ecosystem:
Number of predators and prey	
Quality of habitat	
Food availability	

Finding the Highest Leverage Point

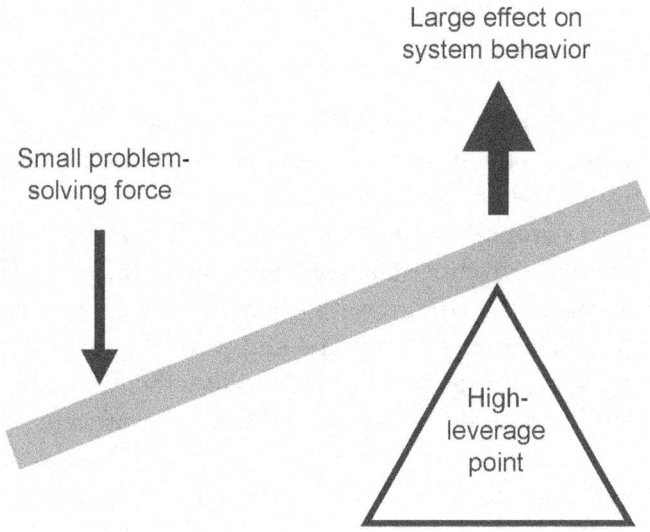

FIGURE 2.3 Concept of a High-leverage Point.

of identifying this difference is key because the only reliable way to find high-leverage points is to first find the root causes. Unfortunately, fearing time delays or costs, people often will choose a solution that only addresses symptoms of a problem. As Peter M. Senge (2006) emphasizes, "If symptomatic solutions are employed as if they are fundamental solutions, the search for fundamental solutions stops and shifting the burden sets in" (p. 110). "Shifting the Burden" is a systems archetype that explains how people merely relieve symptoms instead of curing the illness—how tackling symptoms, rather than fixing fundamental problems, can lead to a further dependence on symptomatic solutions.

For example, many people depend on pesticides that temporarily remove unwanted pests. However, the peer-reviewed study "Pesticides and Soil Invertebrates: A Hazard Assessment" by Gunstone et al. (2001) shows that pesticides widely used in American agriculture pose a serious threat to organisms necessary for healthy soil, biodiversity, and the fight against climate change. The fact that U.S. regulators fail to consider these harms when approving pesticides for use suggests a preference for a symptomatic solution that ultimately "shifts the burden" and creates grave side effects, threatening the long-term health of the soil that feeds us.

Infusing Lessons With a Systems Thinking Approach

When I began to incorporate systems thinking tools into my lesson planning, the payoff was huge: my students began to see the necessity of comprehending system-wide impacts if they were serious about environmental problem-solving. Infusing your lessons with causal loop diagramming can help learners distill their insights into visual systems maps and pave the way for identifying strategic interventions. Consider how the following lesson tasks would benefit from a systems thinking approach:

- In a worksheet created for students, an educator asks students to list facts about an endangered species
- An educator asks students to debate the merits of fossil fuel divestment
- A student is tasked with writing a multi-paragraph essay about the environmental harm of fast fashion

Lesson #1 revision: Accruing facts about an endangered species is a great starting point for learning about a species population that has reached precarious numbers. However, to better assist learners in understanding why a species is in danger, asking

TABLE 2.2 Infusing Lessons With a Systems Thinking Approach

Lesson Task:	Add a Causal Loop Diagram:
List facts about an endangered species	Draw a balancing loop to explain the elements contributing to the endangered Florida manatee population
Debate the merits of fossil fuel divestment	Draw a balancing loop to explain the impact of fossil fuel divestment
Write an essay about the environmental harm of "fast fashion"	Draw causal loops to make sense of the impact of "fast fashion" consumerism in the fashion supply chain

them to apply their knowledge in the form of a diagrammed causal loop offers a formative assessment opportunity. Merely making a list of facts does not elicit thinking about impacts in the way studying a causal loop does. Changes in ecosystems produce domino effects—in recent years, poor water quality has led to reduced seagrass beds, increasing the number of stranded and dead manatees observed on the Atlantic coast of Florida (*Florida Manatee—Marine Mammal Commission*, 2023). To better understand the chain of effects, slightly alter the task directions: ask students to create an infographic that demonstrates both their research and conclusions based on causal looping.

Lesson #2 revision: Most students would not be able to confidently debate the merits of fossil fuel divestment at length without doing some preliminary thinking. As currently stated, this task misses an opportunity to go beyond debating pros and cons. Provide students with an opportunity to explore the causal relationship between decarbonizing the atmosphere and the institutional divestment of assets connected to companies involved in extracting fossil fuels. Instead of debating the pros and cons of divestment per se, ask students to create a balancing loop based on identifying the highest leverage point in the system: putting pressure on companies highly reliant on fossil fuel extraction.

Lesson #3 revision: Assigning an essay about the environmental harm of "fast fashion" without scaffolded opportunities to map the impacts of the fast turnover of clothing is a missed opportunity. As currently stated, the assignment does not appear to promote collaboration with peers as the writer refines their ideas. However, modeling how to trace the impact of the increased consumption of "fast fashion" through reinforcing loops can help the writer conceptualize the causal relationship between the fast turnover of clothing, the emergence of new trends, and the trend-chasing behavior of consumers, ultimately increasing (reinforcing) fast fashion consumption.

Surfacing Mental Models

As students gain confidence with tracing system feedback and identifying high-leverage points, they develop skills that serve more impactful advocacy: helping the public and policy makers compare the benefits of making a change with the benefits of maintaining the status quo. The status quo, or existing state of affairs, might be something that is hard for students to articulate at the onset of this unit, even if they understand what is happening right now is not sustainable. Some journalists suggest it is more apt to say "global boiling" than "global warming" given the extreme weather events being felt in so many places at once. Current news headlines indicate that residents in the Northeast are experiencing dangerous air quality as a result of uncontrollable wildfires, parts of Florida are suffering devastating flooding, and water reservoirs millions of people rely on are at historically low levels. But the business-as-usual attitude I am referencing has to do with the underlying attitudes that overwhelmingly characterize the human relation to the biosphere.

John Goekler (2009) identifies harmful assumptions inhering in the dominant worldview that is the opposite of thinking in systems:

- Constant and unlimited growth in business is not only possible, but essential
- Humans have dominion over the Earth
- Nature is income—resources are free because we "found" them
- If we destroy our environment, we can simply move or invent some new technology to save us
- We can understand the natural world through reductionism: that is, by breaking it down into small parts

"Teaching for the Future: Systems Thinking and Sustainability" (p. 4)

Activity: Internal Threats

At the beginning of your unit, using a "think-write-pair-share" strategy will help learners process their assumptions. Before partners explore Goekler's aforementioned assumptions in dialogue, ask students to write in response to this prompt.
Journal prompt: Choose the mindset assumption you consider the biggest threat to our planet's viability as a home to future generations. Explain why.

It's not uncommon for students to ask about the meaning of "dominion" or express confusion about the phrase "Nature is income." This confusion is a great starting point for getting to the core of the matter—the often unconscious attitudes that drive our harmful actions—and helping them establish common ground about alternative ways of thinking (Kolbert, 2017).

Activity: Demystifying Harmful Assumptions

After students debrief their response with a partner, engage in a whole class discussion while filling in a poster-size chart or a projected screen image containing the five assumptions. Turning this chart into a poster that is displayed for the duration of the unit is a useful visual tool that can be revisited again and again. It can look something like this:

Mindset Assumptions to Combat:

Unlimited growth is the ideal business model
Humans are hierarchically situated above all living things, instead of existing in a web of interrelated ecosystems
Natural resources are available for human consumption
We can just look for another place to inhabit or rely on technology to fix the problems needed to be solved
We can ignore the interconnected web of the natural world

This conversation is an important step in helping students engage in root cause analysis: diagnosing the root of the problem, not just discussing the symptoms. Asking students to debate which mindset assumption is most harmful can elicit insights that help everyone understand the causality of climate change events. See a student exit ticket response:

> The 'Nature is income' mentality makes people think it is okay to destroy entire forests and ecosystems just so they can advance their own business to make more profit. This mindset has been flourishing ever since the Industrial Revolution began and we are now seeing the effect on

rising climate temperature, the decline in animal species, and our melting ice caps.

Educating about feedback loops prepares learners for understanding the danger of anthropocentric thinking, or seeing ourselves at the center of activities. Peter M. Senge (1990/2006) explains the importance of overturning this idea:

> *From the systems perspective, the human actor is part of the feedback process, not standing apart from it. This represents a profound shift in awareness.* It allows us to see how we are continually both influenced by and influencing our reality. It is the shift in awareness so ardently advocated by ecologists in their cries that we see ourselves as part of nature, not separate from nature.
>
> (p. 77–78)

Unpacking Assumptions

The Ladder of Inferences, a valuable tool for reflecting on the origins of belief, is a mental model first described by organizational psychologist Chris Argyris (1990). The ladder is made up of seven rungs that outline the rapid process our minds go through to make inferences and take action in a given situation. According to Argyris, our beliefs emerge from the conclusions we draw from assumptions we hold. "These assumptions form when we attach meaning to information that emerges from our personal experiences" (Walsh & Sattes, 2015). Argyris reasons that the problem with this practice is that our habit, once a belief is formed, is to return to a default mental setting and look for reinforcing data, a mental impulse we know as the logical fallacy of confirmation bias.

Bringing the Ladder of Inferences into the classroom can serve as an effective entry point for scrutinizing one's own tendency to look for data that reinforces preexisting assumptions. One reason systems thinking is a powerful system of approaching problems is the reflective disposition it cultivates. Prior to deciding upon a course of action, students learn that taking stock of their beliefs is part and parcel of figuring out whether they are predisposed or disinclined to seriously weighing a solution held up to scrutiny. A classroom conversation that arose a year ago sheds some light on the Ladder's potential usefulness. My students were struck by the number of Super Bowl television ads devoted to promoting electric vehicles. As I listened to them analyze how celebrities were used to draw attention to newly released EVs—from Arnold Schwarzenegger dressed as Zeus, wielding electrifying lightning bolts, to Mike Myers assuming his persona as Dr. Evil, hatching a new plan for world domination—I considered what kind of impact the ads were making. Seen by roughly 100 million people, the ads no doubt were hugely influential in creating brand awareness. But the suggestion that EVs were on the verge of mass adoption produced a mixed reaction: some students were on board, a few were indifferent, and some looked openly hostile to the prospect.

When uncomfortable moments in the classroom arise, educators might be tempted to steer students away from potentially stormy waters, perhaps unsure if they themselves have the conversational tools to navigate discussions with students who are emotionally charged. In these moments, tools meant to be used to explore the roots of underlying attitudes and assumptions can be a North Star. I introduced the Ladder of Inferences tool by drawing attention to its potential for cultivating awareness about when our beliefs arose. In this instance, I asked my students to reflect with each "rung" of the ladder, as they pondered the origins of their views on electric vehicles. After being given ample time to do so, I invited students to share what they noticed on a voluntary basis. Some were surprised to consider the extent to which their views aligned with the people who raised them; others felt that enthusiasts who championed the use of EVs were elitist, but they could not point out one incident that gave rise to that view. We talked about how stigmas associated with certain environmental solutions (such as switching to an electric vehicle) seem woven into geographical contexts and how a proposed solution may appear to threaten one's values or way of life (a love of classic cars shared with a family member, anxiety about how a shifting car market could impact a relative's auto repair business). Having the opportunity to think *about* the way they think made them more conscious about when and why proffered solutions produced emotional reactions that appear knee-jerk and reflexive. Zaretta Hammond (2015) finds there are five elements of social interaction that activate strong threats and rewards in the brain, thus

TABLE 2.3 Social Interaction Elements That Activate Threats in the Brain by Zaretta Hammond

Element	Description	What's the Threat?
Standing	Standing refers to one's sense of importance relative to others in one's social network or organizational hierarchy (e.g., peers, coworkers, friends, supervisors). It also relates to how one believes others in the group perceive them—negative or positive, competent or incompetent.	The fear that one would be expelled from the "tribe" (such as being fired from the job, evaluated poorly by a supervisor, ostracized by peers because of doing things differently).
Certainty	Certainty refers to one's need for clarity and predictability in a social situation in order to make accurate social moves. It also relates to one's ability to predict what will happen (e.g., routines, cause and effect, action and reaction).	The fear of possibly embarrassing oneself or being unable to know what to do in a given situation. The feelings of being out of control or unable to be safe because of venturing into the unknown.
Control	Control speaks to one's sense of control over their life and the perception that one's behavior can have a positive effect on the outcome of a situation (e.g., getting a promotion, finding a partner) rather than something out of their control having more influence (e.g., class, race, language, or gender).	The fear of someone telling you what to do, where to go, and how to behave that is inconsistent with your values (such as with English-only laws or Jim Crow laws).
Connection	Relatedness focuses on one's sense of connection to and security with another person, one's family, or one's peer group. It also is concerned with whether new people one interacts with are friends or foes.	The fear of being an outsider and excluded. We fear losing important connection with others. People do not want to be out of relationship with others, especially an important peer group.
Equity	Equity refers to having a sense of fair, just, and unbiased exchange between people (e.g., equal opportunity, equivalent pay for equivalent work, the elimination of unearned advantage and disadvantage).	The threat can come when one feels their group (class, geographic, linguistic) is being subjected to unearned disadvantage or someone is receiving unearned advantage. It may also be associated with distancing oneself from unearned advantage.

Source: Hammond, Z. (2015). *Culturally responsive teaching & the brain: Promoting authentic engagement and rigor among culturally and linguistically diverse students.* Corwin.

influencing how we behave in given situations: *standing, certainty, connection, control,* and *equity* (Figure 1.6). Consider which threats surface during these types of classroom conversations.

Taking the time to create classroom conditions where learners can interact constructively with those with opposing views is worth the effort. Researchers Joseph Kahne and Benjamin Bowyer (2017) find, as a result of the increasingly politically polarized environment characterizing the United States in recent years, "judgments of truth claims are often shaped more by whether or not individuals' prior perspectives on the issues align with the claims than by how well informed the individuals are or their capacities to reason." Using reflective tools like the Ladder of Inferences is an example of emotional scaffolding: the use of metaphors, analogies, and narrative to frame a lesson in a way that improves the emotional dimension of students' encounter with the subject matter (Rosiek, 2003; Wolfe, 2019). Emotional scaffolds construct an environment that helps students discover themselves in relation to new ideas and open up conversations that are not just about finished thoughts.

Activity: Using the Ladder of Inferences

Invite students to reflect on their beliefs regarding an environmental solution they are familiar with. If they struggle to begin,

TABLE 2.4 Ladder Rung Student Reflection

Ladder Rung Mental Process	Student Reflection
7. I take actions based on my beliefs.	
6. I adopt beliefs.	
5. I draw conclusions.	
4. I make assumptions.	
3. I add meaning from personal experiences.	
2. I select "data" on which to focus.	
1. I have experiences and make observations that give me data about the world.	

Source: Walsh, J. A., & Sattes, B. D. (2015). *Questioning for classroom discussion: Purposeful speaking, engaged listening, deep thinking*. ASCD.

you might suggest they think of images they are familiar with through mass media or consulting their memories of exposure to the topic.

Activity: Engaging in Reflection

In this work of developing holistic thinking and creating pathways to climate stewardship, it is important to think about the first and early messages you received that inform and impact the educator you are today. Read the following reflection questions and record your thoughts and experiences in the space provided.

Using the Ladder of Inferences as a reference, what personal experiences have you had that inform and shape your perception of climate stewardship? What conclusions have you drawn as a result of these experiences?

As a result of your beliefs about climate stewardship, what actions have you taken (or not taken)?

Just as you and the educator teaching next door might answer these questions very differently, so will the students in front of you. Reflecting on the formative influences shaping your beliefs about what it means to sustainably coexist on this planet will help you create a space where students can authentically and seriously reflect on early messages they received about what it means to take care of our planet home.

The Role of Logical Fallacies

Using the Ladder of Inferences in conjunction with studying common logical fallacies can be meaningful practice in preparation for systems mapping towards solutions. Illuminating mental barriers such as heuristics—cognitive shortcuts by which people generate judgments and make decisions without having to consider all the relevant information—as well as biases that hamper human observation can empower students to actively counteract them (Peer & Gamliel, 2013). For example, when my students debriefed what they learned while using the Ladder of Inferences to explore the origins of their beliefs, some offered revealing anecdotes about when their stance on electric cars arose.

One student had watched a YouTube video where an unnamed driver had attempted to charge a rented EV at a charging station, only to be greeted by the sight of multiple out-of-order charging stalls. The video narrator, who began the video by professing a desire to find out the range possibilities of the rental car firsthand, announced this was only one of several failed charging attempts, confirming what he already suspected about unreliable range estimates. Upon viewing, my student surmised this charging experience was typical and identified the video as the source for his disinclination to take electric cars seriously as a climate solution.

The more we learn about the data students select from which they make assumptions and draw conclusions, the better educators can become at designing lessons that help students sift through the vastly expanded opportunities for circulation of both information and misinformation (Kahne & Bowyer, 2017). One way to prepare them is to acquaint learners with common biases and fallacies frequently encountered in conversations about solutions. For example, as a result of hearing the group of students reflect, I realized that frontloading a solutions unit with a lesson about logical fallacies enabled learners to make their own connections between their assumptions and messages they have absorbed. This approach has three critical advantages. One is we help learners develop media literacy at a time when many are not aware of how deeply partisan biases affect their receptiveness to ideas, or have sufficient practice with identifying criteria for valid news sources (Weissbourd et al., 2023). Two is we help them recognize the verbal tactics climate offenders often use to shift blame, enabling learners to become more savvy consumers of media. Three is we help them assess how solutions are being rhetorically framed in conversations: how verbal cues potentially create buy-in or mistrust in the audience.

Table 2.5 presents definitions for some of the logical fallacies commonly encountered in conversations about climate solutions.

Once learners begin to practice identifying examples of fallacious statements, they start to notice fallacies exist ubiquitously in surrounding conversations. In the example of the YouTube video anecdote shared by my student, the video's narrator

TABLE 2.5 Common Logical Fallacies

Either/Or fallacy: occurs when someone incorrectly presents two possible options as the only ones	Hasty generalization fallacy: occurs when you generalize a singular experience or small sample size
Straw Man fallacy: occurs when someone distorts an opposing stance in order to make it easier to attack	Cherry picking fallacy: occurs when someone focuses only on evidence that supports their stance, while ignoring evidence that contradicts it
False cause fallacy: occurs when the link between premises and conclusions depends on some imagined causal connection that probably does not exist	Confirmation bias fallacy: the tendency to look for information that corroborates what we already believe
Appeal to tradition fallacy: occurs when a conclusion is supported solely because it has long been held to be true or superior	Default bias fallacy: automatically favoring or accepting a situation simply because it exists right now

engages in confirmation bias: he was looking for evidence of his belief that EV charging stations could not be relied upon when charging was necessary. As a result of interpreting the narrator's experience as the rule and not the exception, my student engaged in hasty generalization, taking for granted that this one example exemplified the norm. Providing examples of fallacious reasoning helps learners recognize when they themselves are prone to it.

Case Study: Plastic Waste

How an educator defines the parameters for approaching problem-solving profoundly influences the potential for real-world application. An important first step in setting up the parameters is helping students scope out existing solutions already under scrutiny. The invitation to posit the best solution in the absence of meaningful skill practice, however, only allows students to skim the surface. In order for students to have something substantive to say, they need opportunities to determine the best course of action given the weighing of possible options. This means exposing students to what is being said about the problem and assessing why one solution is being championed versus another. We can model what this weighing looks like by showing students what we notice about what is being said about a solution in the rhetorical landscape: how it is framed and what is being emphasized. As Jennifer Fletcher cogently advises in her book *Teaching Arguments* (2015): "We might need to do a little reconnaissance listening before we're ready to join the conversation" (p. 24).

Educators can set up the conditions for this kind of reconnaissance by modeling inquiry that brings the problem into focus. Take an example from Xiye Bastida's 2020 TED Talk, "If you adults won't save the world, we will." Written as a letter to her abuelita, her speech explains her reasoning for participating in climate strikes. As she weaves references to the example set by her family members, and to the influence of Greta Thunberg, who called for the first global climate strike, she describes her origin story as a climate activist. She recalls joining her high school's environmental club and observing a type of environmentalism that places its focus on what an individual can do, "one that blames the consumer for the climate crisis and preaches that temperatures are going up because we forgot to bring a reusable bag to the store."

Bastida's critique gets to the heart of a flawed approach to what we can do to protect our planet—focusing chiefly on the individual efforts of a consumer. This focus is a symptom of a purposeful campaign to put the onus of responsibility on the individual consumer rather than on corporations accountable for mass producing so much wasteful single-use packaging. She is diagnosing an example of what is called "the false cause fallacy": relying on an imagined causal connection between rising temperatures and failing to use a reusable bag for store purchases to make an argument. In her TED talk, she goes on to explain that focusing chiefly on what the individual can do is at odds with the wisdom her grandparents have shared with her about the importance of taking collective action. Ultimately, she persuades the environmental club members to write letters to politicians to ban soft plastics, underscoring the idea that *just* talking about recycling was insufficient. These conversations are typical of traditional environmental education efforts, which "tend to focus on teaching students *about* climate change—a passive approach that fails to capitalize on students' capacity to make a meaningful impact in their communities *now* while inspiring their commitment to future environmental work" (Hausburg & Herrmann, 2023).

Thinking like a climate steward includes diagnosing root causes and possible leverage points: the more opportunities teachers provide for identifying them, the more adept students will become in posing counterstories—ideas that transform harmful mindsets and behaviors. Many people attempt to address the plastic waste problem by using reusable bags when shopping (what would be considered a low-leverage point in terms of impacting a system). Persuading businesses to curb their single-use packaging, though, would unquestionably apply pressure as a high-leverage point and tackle the root cause of the problem.

Introducing the ripple effect tool will help learners distill their insights about the relative strengths and weaknesses of the solutions they are contemplating because it gives them a chance to assess consequences in a visual, accessible manner. Planning for student groupings for this activity can assist learners in seeing how chosen solutions will potentially play out, even giving them a chance to hypothesize whether a choice will increase the effectiveness of the whole system over time or merely meet the needs of a short-term time frame. When educators provide learners with opportunities to identify high-leverage interventions, students gain valuable experience in thinking about how to focus limited resources for maximum improvement.

There is a famous example of unintended consequences using the ripple effect tool. In an attempt to control the population of venomous cobras that were plaguing the residents of Delhi, the British Colonial Government offered a bounty to be paid for every dead cobra brought to the administration officials. While the policy initially appeared successful, it was an illusion: snake catchers realized that breeding and then killing cobras made for a lucrative source of income. Government officials were puzzled by the number of payments being made despite the number of cobras still seen around the city. Once the secret breeding sites

Climate Stewards as Systems Thinkers ♦ 67

were discovered, the government abandoned the bounty policy. As a consequence, the breeders released the remaining cobras into the city, compounding the initial problem (Systems Thinking and the Cobra Effect—Our World, n.d.).

The ripple effect tool illustrates the consequences of initial actions, often shedding light on possible negative, future outcomes. Asking students to use this systems thinking tool to weigh different solutions is an effective way to generate collective brainstorming. Figure 2.4 displays a student group's mulling of how a specific environmental solution would play out:

As my students contemplated the frequency of extreme weather events, their thoughts turned to fire. The fire scars created by uncontrolled wildfires create a domino effect of problems.

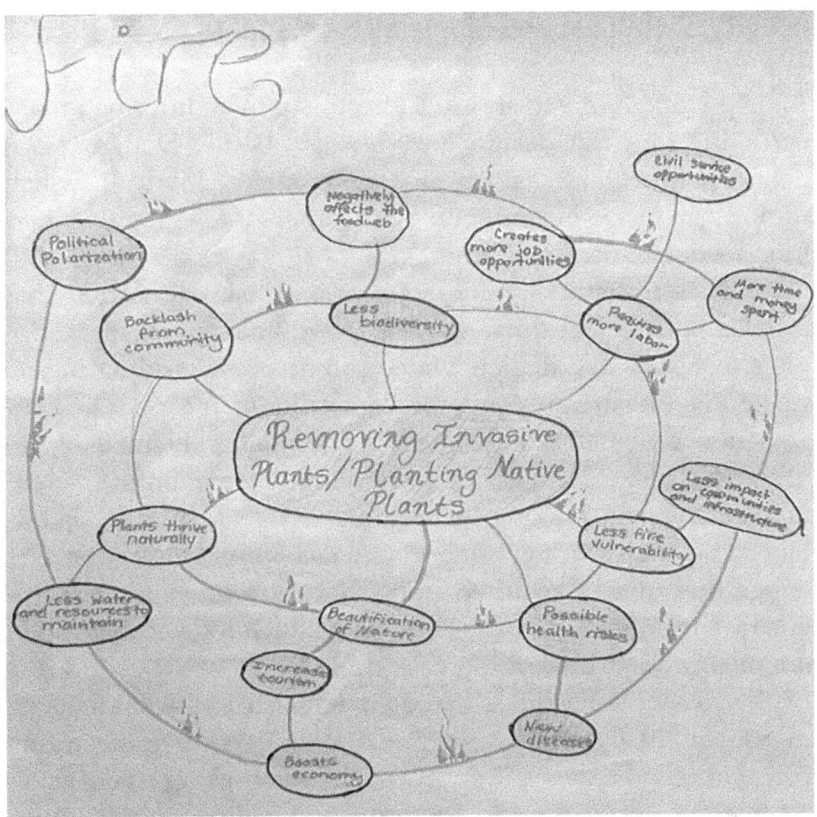

FIGURE 2.4 Student Example of Ripple Effect Mapping. Photograph by the Author.

While weighing possible solutions, the group kept circling back to the prospect of planting native plants and removing invasive plants. As they discussed the upside (native plants thrive naturally, ultimately requiring less water and fewer resources for maintenance), they were mindful of perceived downsides (more labor required). Yet, more discussion elicited the realization that this outcome could lead to advantageous ones (more civil service job opportunities). This systems thinking tool encourages thinking across domains and at different scales, much like problem-solving in real-world contexts does. The ripple effect tool can be a student-made visual that they can return to as often as needed, especially as their thinking about a chosen "best" solution evolves with their research investigation.

Activity: WebQuest and Ripple Effect Mapping

To help students create a ripple effect map in an informed manner, educators can design a WebQuest. Bernie Dodge (1995) defines a WebQuest as "an inquiry-oriented activity in which some or all of the information that learners interact with comes from resources on the Internet."

Educators gather online sources and ask a question that focuses student scrutiny, e.g., how would you create a ripple effect map that depicts viable solutions for tackling plastic waste? To prepare students for their culminating podcast conversation, designed around an environmental problem, they will need practice in synthesizing important information from credible sources and making connections between specific actions taken and the "rippling" impacts. A WebQuest helps educators gauge how newly acquired information transforms the learner's understanding (March, 2003). Table 2.6 shows a collection of online sources to anchor this case study:

Inspired by Bastida's critique, I instruct my students to investigate how our biggest climate offenders—manufacturing corporations—strategically divert attention from themselves by placing focus on the individual behavior of consumers (Dunaway). After World War II, companies shifted to lighter

TABLE 2.6 Example of Gathered WebQuest Sources

Building Background Knowledge: Plastic Waste
CBC/Radio-Canada. (2019, September 27). *Tracking your plastic: Exposing recycling myths*. YouTube, uploaded by CBC News. https://www.youtube.com/watch?v=c8aVYb-a7Uw
Liboiron, M. (2018, December 21). How plastic is a function of colonialism. *Teen Vogue*. https://www.teenvogue.com/story/how-plastic-is-a-function-of-colonialism
McLeod, M., et al. (2021, July 2). The global threat from plastic pollution. *Science, 373*(6550), 61–65. https://www.science.org/doi/full/10.1126/science.abg5433
Sedaghat, L. (2018, April 4). 7 things you didn't know about plastic. *National Geographic*. https://blog.nationalgeographic.org/2018/04/04/7-things-you-didnt-know-about-plastic-and-recycling/
Borrelle, S. B., et al. (2020, September). Predicted growth in plastic waste exceeds efforts to mitigate plastic pollution. *Science, 369*(6510). https://www.science.org/doi/10.1126/science.aba3656

Linear Economy

FIGURE 2.5 Example of a Linear Economy.

single-use bottles to reduce costs. While this move kept costs down, corporations did nothing to manage the environmental impact of their trash on communities. This behavior contributes to what we call a linear economy.

What ensues is what we call "throwaway culture," where consumer products are used once only, from disposable packaging. It can be illuminating for students to study the 1953 Keep America Beautiful campaign: a public advertising campaign created by the packaging industry and large companies such as PepsiCo and Coca-Cola. The goal of the campaign was to shift the responsibility for accumulating litter in communities to consumers and away from polluting corporations. In fact, it was a writer at the American Advertising Council who six years earlier coined the term "litterbug"—someone who threw their trash in places other than trash cans (*Keep America beautiful litter study ignores corporate blame for plastic pollution—Greenpeace USA, 2021*). This is the same Ad Council who created public service announcements for Keep America Beautiful. Greenwashing campaigns such as Keep America Beautiful do not get to the root cause of the problem: the waste occurring on the production side (Dunaway, 2017). These campaigns may include images of nature or use terms like "eco-friendly," but greenwashing marketing tactics merely create the image of environmental friendliness. Real systemic solutions—where high-leverage points can be found—are subsumed into discussions of what the individual consumer should be doing.

Box 2.1: Standalone Quotation

The great political transition of the past 50 years, driven by corporate marketing, has been a shift from addressing our problems collectively to addressing them individually.

George Monbiot, 2021

Learning From Dystopian Fiction: How We Destabilize the Status Quo

Selecting a text that immerses learners in a fictional world has been a key means of engaging students in the practice of becoming climate stewards who are systems thinkers. While it may seem more logical to pair exploration of key concepts with non-fiction texts, written by climate leaders directly tackling the harmful impacts of climate change, the readers of dystopian fiction often find answers for how and why negative social forces have created destructive ills. Much dystopian fiction relies on the process of defamiliarization: a technique that presents familiar issues in unconventional settings, which can elicit new insights when the familiar is no longer perceived as such (Soares, 2020). Readers are provided with "a counter-narrative that creates a level of suspicion about the surface of the so-called realities they experience," opening up alternative ways of viewing contemporary problems (Han et al., 2018). This expanded mental terrain is fertile ground for planting seeds—suggestions for how we might meet these challenges in our own contemporary circumstances. Stroh (2015) explains that systems stories are driven by a specific question: "Why are people unable to solve a chronic, complex problem or achieve a meaningful goal—often despite their best efforts?" (p. 45). Fiction, by virtue of its ability to hold a mirror up to human nature in an oblique manner, can allow us to see our own culpability in perpetuating these problems. One way to help students become climate stewards is to model how reading dystopian fiction enhances our ability to both problem-spot and problem-solve.

Case Study: Lauren Olamina in *Parable of the Sower*

Reading Octavia Butler's *Parable of the Sower* in the present tense is a slightly eerie experience; so much of what Butler has presented in her fictional novel set in the 2020s is happening: uncontrolled fires, resource depletion, and rising sea levels. The novel appeared on the *New York Times* bestseller list in 2020, 23 years after its initial publication. Butler's prescient storytelling is remarkable for how it shows the convergence of environmental and societal disintegration, as revealed through the character of Lauren's narration:

> The cost of water has gone up again. And I heard on the news today that more water peddlers are being killed. Peddlers sell water to squatters and the street poor—and to people who've managed to hold on to their homes, but not to pay their utility bills. Peddlers are being found with their throats cut and their money and their handtrucks stolen. Dad says water now costs several times as much as gasoline. But, except for arsonists and the rich, most people have given up buying gasoline.
>
> (p. 17–18)

Part of what distinguishes Lauren as a character is her willingness to acknowledge and anticipate problems that have begun to snowball—a mental disposition that counters the perspective of the adults in her community, a walled neighborhood in Robledo, California, where many embrace the ideas of President Donner, a politician who promises to return things to normal. In marked contrast, Lauren strives to adjust to uncertainty, as she recognizes this push for a return to normalcy provides insidious cover for the brutal violence and the unequal distribution of resources that have been normalized.

When we discuss Butler's novel, my students frequently describe Lauren as someone who is able to "think outside the box." I encourage them to unpack this cliché and dig deeper. It's not unusual for the protagonist of a dystopian novel to have a perspective that diverges from the norm—this divergent outlook frequently winnows a path for this character to push back against the controlling forces that perpetuate the bleak societal circumstances typical of a dystopian setting. Reading my students' journal entries taught me that Lauren's willingness to find solutions modeled a specific type of leadership. By unpacking the cliché of "thinking outside the box," we noticed that a key feature of Lauren's emerging leadership was an ability to see the flaws and the need for improvements in systems.

Through applying the "iceberg model" map sometimes used by systems thinkers, we can see how Lauren is able to look beneath the surface level, and observe how events are parts of larger patterns. Above the water line, we identify the event level that is purely reactive. At the patterns of events level, we can see the behavior that our systems create over time. At the systemic structure level, we see the political, economic, and social structures that create these patterns. Finally, the mental model (paradigm) level reveals our beliefs about how the world works—a critical leverage point for envisioning new results:

- Events (What just happened?)
- Trends/Patterns (What trends have been there over time?)

- Underlying Structures (What are the relationships between the parts?)
- Mental Models (What beliefs keep the system in place?)

Activity: Intergenerational Awareness

Thieves, increasingly desperate and bold, crowbarred their way into the Robledo community by disabling burglar alarms. In their haste to get away, they knocked over and killed an elderly woman, Mrs. Quintanilla. Processing and reacting to this event is a huge turning point for the members of the walled community—Lauren's reaction is a clear divergence from her friend's hope of finding a better life in the coastal city of Olivar.

As the community struggles unsuccessfully to fortify itself in the hope of forestalling an even deadlier invasion, Lauren considers what kind of example the adults are modeling. The question of "What do we inherit from our elders?" can be a helpful entry point for considering differing reactions to the event and the intergenerational dynamics that inform Lauren's growth as a character. Building in notebook time for addressing journal prompts can help students better understand the barriers to the characters' collective growth.

Journal Prompt: What do we inherit from our elders?

One student's journal writing sums up the contrast beautifully: "All the adults in Lauren's community choose to deny the reality of their neighborhood collapsing; the children are also inheriting this denial."

Anticipating Patterns

My students note that Lauren foresaw the opportunistic behavior of the "land sharks." One exit ticket response articulates how her coming of age story is mapped by her developing foresight:

> Lauren uses a metaphor illustrating how she anticipates dangers. The sharks represent the people beyond the border who prevent them from leaving their community. She introduces a potential emerging conflict when she states it's only a matter of time before the "sharks" get hungry enough—possibly foreshadowing an attack.

As the number of thefts within the neighborhood walls increases, community members seriously contemplate the prospect of applying to join a nearby privatized city: Olivar. This juncture of the novel reveals how people still opt for the status quo, even in pursuit of a better life. While working in literature circles, my students diagnosed the problematic aspects of entering into a contractual agreement with KSF, the company that has taken over the small, coastal city. Though understanding how receiving the "protection" of KSF security would be a welcome contrast to the recurring threat of violent theft, labor laws are incredibly lax in President Donner's America. Lauren is able to discern the "old company-town trick": if you get people into debt, you can lock them into a contract and work them harder.

The ability to identify patterns as Lauren does is critical for mapping out alternatives to the default habits of thinking that offer little room for hope. Instead of falling back on the same extractive models of labor, Lauren imagines the possibility of heading north and forging a new community.

Underlying Structures

Systems thinking raises questions about the relationship between the parts—the political, economic, and social structures—that underpin the patterns of behavior these systems create over time. When Lauren's friend, Joanne, asks her if privatized cities will become commonplace, Lauren is candid in responding: "When people like those in Olivar beg to sell themselves, our surviving cities are bound to wind up the economic colonies of whoever can afford to buy them." In discussion groups, my students thought about the crumbling infrastructure of Lauren's

TABLE 2.7 Making Connections Through Root Cause Analysis

Problem Symptom	Problem Root Cause
Water costs more than food.	Because drinkable water is scarce, a handful of water companies have formed a monopoly.
Wages don't match the cost of living.	More and more people have resorted to violent theft due to few existing employment options.
Most adults regard voting as a waste of time.	People have lost faith in the government's capacity to improve societal conditions.
Few children attend school or receive any instruction.	The children in Lauren's walled community are illiterate.

society and considered the relationship between these problems and their root causes as seen in the novel. Table 2.7 documents the connections they made between identified symptoms and root causes.

Activity: Avoiding the Quick Fix

While fostering systems thinking skills, offer frequent opportunities for students to consider the difference between foundational solutions and short-term, Band-Aid fixes.

> Journal Prompt: How do we see evidence of characters choosing Band-Aid fixes?

My student's journal entry articulates the problem of addressing the symptom and not the root cause:

> Moving to Olivar is a short-term solution to the fear of home invasion. The community members who have applied to join the city are not thinking things through. Sometimes, city structures within Olivar crumble and fall into the ocean, since much of the city is already steeped in saltwater.

This student does a great job of communicating the relationship between poor city planning and extreme coastline erosion. What certain community members are willing to do to "get by" means putting down roots in a city that lacks a sustainable infrastructure.

New Paradigm of Thinking

The willingness to accept change and adapt is the cornerstone of Lauren's Earthseed writings and the "Acorn" community she establishes. Throughout the novel, she has observed the destructive beliefs that keep a system of unjust resource distribution in place:

- You can maintain power and influence through violence
- Wealth accumulation occurs through the hoarding of resources
- Surrendering freedom is worth it if your safety is guaranteed
- Holding onto the past will establish a sense of normalcy

In contrast to these beliefs that perpetuate mistrust, Lauren's Earthseed community is focused on linking personal well-being, community well-being, and planetary well-being. This level of the iceberg model can help students interrogate the stories the characters tell themselves and the model of leadership offered through Lauren's restoration of community.

Activity: Iceberg Model Mapping With Desalination

Discuss this mini-case example from the novel: Olivar has a solar-powered desalination plant that provides its people with a dependable supply of water. Desalination is an energy-intensive process of removing salt from seawater. Use the iceberg model levels to explore the environmental impact of this process.

TABLE 2.8 Desalination: A Good Solution to Water Scarcity?

Iceberg Level:	Student Reflection:
Events (What just happened?)	
Trends/Patterns (What trends have been there over time?)	
Underlying Structures (What are the relationships between the parts?)	
Mental Models (What beliefs keep the system in place?)	

From Awareness to Knowledge to Action: Creating a Podcast Episode

This chapter has presented several tools that can help learners destabilize "status quo" thinking: their own habits of mind and those of a potential podcast audience. Forming a question about an environmental problem and possible solutions that will guide their research and podcast development is the first step. In the face of a looming project, however, students can get stuck. Once they have established a line of inquiry, they will need guidance in planning the structure of their podcast conversation. To help them plot their talking points, offer these considerations as mental footholds for group planning:

- Investigate a solution in terms of perception versus reality: by examining what is said about an environmental solution, podcast participants can mull where perceived and actual benefits actually align
- Explore consequences of environmental solutions through system feedback and ripple effects: use systems thinking tools such as causal feedback loops and the ripple effect tool to weigh how solutions play out
- Zoom in on key terms: signpost the usefulness of staging dialogue to demystify terms that might be currently abstract in the listener's head

TABLE 2.9 Podcast Planning Tool: From Topic to Story

Phase	Entry Points
Develop Line of Inquiry	• Pose a question about current environmental problem • Research barriers to implementing a long-term solution
Convey Complexity of System Dynamics	• Explore system feedback with causal loop diagramming • Prioritize investigation goals
Surface Mental Models	• Explain how the problem and potential solutions are rhetorically "framed" • Reflect on underlying assumptions that perpetuate the problem
Choose Solution	• Articulate the root cause versus the symptoms of the problem • Assess the problem through different scales and in different contexts
Engage Key Stakeholders	• Identify policymakers and decision-makers that operate at highest leverage point in a system • Choose specific actions that can effect change at that point in a system
Plan Structure of Podcast Conversation	• Create and revise talking points • Support arguments and ideas with specific examples

- Reflect on one's own role in exacerbating the problem: podcast participants should consider the part of the system in which they play a role, as well as attend to any biases surfaced when thinking about their current impressions of the problem and potential solutions.

Studying the Conversation Moves of Podcasters

Asking students to create a podcast episode is a useful assessment of systems thinking competency. Like Xiye Bastida and Lauren Olamina, students should diagnose environmental problems by addressing root causes, not merely symptoms. Educators can provide podcast mentor texts that help students articulate the value of developing hindsight and foresight—"causal analysis and prospection"—in gauging the likelihood of a solution's long-term success and hypothesizing why it will not (Bracher, 2022, p. 30). Rebekah O'Dell and Allison Marchetti (2021) offer guiding questions for studying structure that can steer would-be podcasters in a promising direction:

> What are the parts and pieces of this text?
> What does the writer do to begin?
> What does the writer do to end?
> How does the writer divide or chunk the writing?
> What elements of this writer's structure might you borrow for your own piece? p. 60

Students can study transcripts of mentor text podcasts, annotating their noticings about key craft moves and patterns.

TABLE 2.10 Mentor Text Noticings About Podcasts

Mentor Texts	Make Noticings	Look for Patterns
Throughline, "The Litter Myth" (September 15, 2019)	Asks the well-placed question: "How did the responsibility for keeping the environment clean fall on us, the consumers, rather than the companies that make the waste?"	Many podcast episodes devoted to discussing sustainability topics voice a key question midway that offers a major idea.
The Takeaway, "The Ethics of 'Fast Fashion'" (November 5, 2022)	Offers an actionable step: . . . all you have to look for is that social responsibility page and look for the place where they say, "All of our garment workers make fair and living wages."	Many podcast episodes devoted to discussing sustainability topics offer insight about what the listener can do to make a more ethical choice.
TILclimate Podcast, "TIL about Electric Cars" (August 25, 2022)	Establishes the stakes: "About 20% of all greenhouse gas emissions in the United States come from what we think of as cars."	Many podcast episodes devoted to discussing sustainability topics include an expert that establishes why the topic is important in the first place.

TABLE 2.11 Systems Thinking Podcast Rubric

	4	3	2	1
Produce podcast that conveys complex dynamics of an environmental problem	Podcast clearly conveys complex dynamics of an environmental problem.	Podcast conveys complex dynamics of an environmental problem.	Podcast conveys at least one system variable contributing to an environmental problem.	Podcast insufficiently or unclearly conveys understanding of an environmental problem.
Explain how causal loop diagramming elicits insights about how the problem persists	Podcast episode clearly explains how causal loops diagramming elicits insights about how the problem persists.	Podcast episode explains how causal loops diagramming elicits insights about how the problem persists.	Podcast episode explains how causal loops diagramming elicits at least one insight about how the problem persists.	Podcast episode insufficiently explains how causal loops diagramming elicits insights about how the problem persists.
Engage in reflection about mental models that inform understanding of the problem and solutions	Podcast participants clearly articulate and reflect on mental models that inform understanding of the topic.	Podcast participants articulate and reflect on mental models that inform understanding of the topic.	Podcast participants articulate and reflect on at least one mental model that informs understanding of the topic.	Podcast participants insufficiently articulate and reflect on mental models that inform understanding of the topic.

	4	3	2	1
Create a sequence of talking points that build awareness, develop knowledge, and advise how to take action	Sequence of talking points clearly build awareness, develop knowledge, and advise how to take action.	Sequence of talking points build awareness, develop knowledge, and advise how to take action.	Sequence of talking points somewhat build awareness, develop knowledge, and advise how to take action.	Sequence of talking points insufficiently build awareness, develop knowledge, and advise how to take action.
Delineate how listeners can effect change based on recommended actions operating at highest leverage point in a system	Podcast talking points effectively delineate how listeners can take action operating at highest leverage point in a system.	Podcast talking points delineate how listeners can take action operating at highest leverage point in a system.	At least one podcast talking point delineates how listeners can take action operating at highest leverage point in a system.	Podcast talking points insufficiently delineate how listeners can take action.

Their talking points should encompass the highs and lows of their investigative journey, especially when discussing finding high-leverage points that emerged from identifying the root causes of environmental problems.

Conclusion

Building a school culture where environmental literacy flourishes means re-envisioning what student learning looks like. To move away from a curriculum that perpetuates classroom conditions of thinking and working in independent "silos," lesson design in every discipline needs to offer practical steps for weighing collective understanding of an environmental problem. When we ask students to engage in problem-solving, they might stall as they contemplate the "doom and gloom" messaging typically associated with the effects of climate change. System thinking offers a perspective and a language for describing and understanding the forces and interrelationships that shape the behavior of systems. The diagram tools associated with systems thinking offer a visual means of communication that can convey the complexity of a problem and serve as a memory aid.

Next Steps

The following questions can help learners continue to practice their skills as systems thinkers and translate perception of problems into meaningful action.

Think of a workplace or community problem. How can creating a causal loop diagram develop awareness about how the problem appears in that system?

How does the way the media "frame" an environmental problem create additional barriers to solving it?

Think about your own role in exacerbating or perpetuating a problem. What is the highest leverage for this situation at the level at which you are operating?

A Peek Into Their Practice

Vermicomposting: a process that relies on earthworms and other microorganisms to help stabilize active organic materials and convert them to a valuable soil amendment and source of plant nutrients.

Erica Schatz teaches at Taper Avenue Elementary School in San Pedro, CA. Having observed worm bins at the nearby Christensen Science Center, she thought a vermicomposting project would be a fun, interactive, hands-on way for her fourth-grade students to learn about decomposers. Erica researched the "best worm for the job" and discovered Red Wigglers (*Eisenia fetida*) were hearty decomposers who ate a lot and processed food quickly. Students studied the anatomy of worms and considered the impact of food that is simply wasted and thrown out to rot, putting more methane in the atmosphere. They studied the components of a worm bin, considering what percentage is green waste and what percentage is brown waste. Green waste usually came from the students' leftover breakfast fruits while brown waste came from dried leaves, twigs, and shredded newspaper pages.

There was some trial and error involved, as Erica and her students learned how to maintain the correct amount of moisture in the bins. They were collecting much more food than they could decompose, prompting Erica to add a nearby compost bin, from which they could pull from if necessary. A new challenge arose: making sure fruit flies did not get into the compost bin!

Erica invited her students to take stock of the food waste in their own lives, amplifying the effect of observing firsthand how food waste can be minimized during composting. Students would express delight in finding an egg or seeing a baby worm, and remind each other to take care when handling the bin contents ("Don't be rough!"). While a few students began the project seeing the worms as "gross" creepy-crawlies, by the end every student came to feel protective towards them as they saw the worms at different stages of their life cycle ("Look at this tiny egg!"). When students from other classes came near the worm bins, Erica's students would offer warnings to be careful, demonstrating how their capacity to be empathetic stewards had been strengthened by their project experience.

Erica allowed caregivers to witness the progress of the project by posting pictures on ClassDojo. Some parents asked if they could take some worms home, expressing their interest in composting at home and putting worms in their gardens. Her students became more aware of their food consumption habits in the school cafeteria, remarking they realized they didn't have to take food out of rote habit. If they didn't eat the food they grabbed, they asked if they could add it to the compost.

Erica's students wrote about the part of the project they wished to highlight, and these pages were assembled into a book, published by Studentreasures Publishing. I am sure you feel inspired as I do: to help my students learn how food waste can be minimized through decomposition and to create early opportunities for students to see themselves as published writers, finding audiences for their words. Erica offered students the choice of what they wanted to write about:

Erica's prompt: Write one page on anything you have learned about the vermicomposting process that you would like to share with others. It can be about the parts of a worm bin, the anatomy of a red wiggler, how food waste turns into methane, or even describing what "worm castings" are! Have fun and share your knowledge! Be prepared to add an illustration to go with your writing.

Guiding Questions: What are decomposers? What are greenhouse gases?

Learning Standards:

Next Generation Science Standards—Earth and Human Activity:

4-ESS3-2. Generate and compare multiple solutions to reduce the impacts of natural Earth processes on humans.

ELA/Literacy:

W.4.7 Conduct short research projects that build knowledge through investigation of different aspects of a topic.

W.4.8 Recall relevant information from experiences or gather relevant information from print and digital sources; take notes, paraphrase, and categorize information, and provide a list of sources.

Math:

MP. 2 Reason abstractly and quantitatively.

Connections to Crosscutting Concepts:

Systems and System Models—Recycling in Natural Systems
Energy and Matter—Conversion of Organic Materials
Structure and Function—Anatomy of Decomposers

References

Abdelfatah, R. (Host). (2019, September 5). The litter myth [Audio podcast episode]. In *Throughline*. NPR. https://www.npr.org/transcripts/757539617

Argyris, C. (1990). *Overcoming organizational defenses: Facilitating organizationa learning*. Allyn and Bacon.

Bastida, X. (2020). *If you adults won't save the world, we will* [Video]. TED Conferences. https://www.ted.com/talks/xiye_bastida_if_you_adults_won_t_save_the_world_we_will?language=en

Borrelle, S. B., Ringma, J., Law, K. L., Monnahan, C. C., Lebreton, L., McGivern, A., Murphy, E., Jambeck, J., Leonard, G. H., Hilleary, M. A., Eriksen, M., Possingham, H. P., De Frond, H., Gerber, L. R., Polidoro, B., Tahir, A., Bernard, M., Mallos, N., Barnes, M., & Rochman, C. M. (2020, September). Predicted growth in plastic waste exceeds efforts to mitigate plastic pollution. *Science*, *369*(6510). https://www.science.org/doi/10.1126/science.aba3656

Bracher, M. (2022). *Literature, social work, and global justice: Developing systems thinking through literary study*. Routledge.

Butler, O. (2019). *Parable of the sower*. Grand Central Publishing. (Original work published 1993)

CBC/Radio-Canada. (2019, September 27). *Tracking your plastic: Exposing recycling myths*. YouTube, uploaded by CBC News. https://www.youtube.com/watch?v=c8aVYb-a7Uw

Dodge, B. (1995). *Some thoughts on WebQuests* [Online]. http://edweb.sdsu.edu/courses/edtec596/about_webquests.html

Dunaway, F. (2017, November 21). The 'crying Indian' ad that fooled the environmental movement. *The Chicago Tribune*.

Fisher, L. H. (Host). (2022, August 25). TIL about electric cars [Audio podcast episode]. In *TILclimate podcast*. MIT Climate Portal. https://climate.mit.edu/podcasts/til-about-electric-cars

Fletcher, J. (2015). *Teaching arguments: Rhetorical comprehension, critique, and response*. Stenhouse.

Florida Manatee. (2023, May 26). Marine Mammal Commission. https://www.mmc.gov/priority-topics/species-of-concern/florida-manatee/

Goekler, J. (2009). Teaching for the future: Systems thinking and sustainability. In T. Grant & G. Littlejohn (Eds.), *Teaching green: The high school years* (pp. 2–9). New Society Publishers.

Gunstone, T., Cornelisse, T. M., Klein, K., Dubey, A., & Donley, N. (2021). Pesticides and soil invertebrates: A hazard assessment. *Frontiers in Environmental Science*, *9*. https://doi.org/10.3389/fenvs.2021.643847

Hammond, Z. (2015). *Culturally responsive teaching & the brain: Promoting authentic engagement and rigor among culturally and linguistically diverse students*. Corwin.

Han, J. J., Triplett, C. C., & Anthony, A. G. (2018). *World gone awry: Essays on dystopian fiction*. McFarland.

Harris-Perry, M. (Host). (2022, November 5). The ethics of "fast fashion" [Audio podcast episode]. In *The takeaway*. WNYCStudios. https://www.wnycstudios.org/podcasts/takeaway/segments/ethics-fast-fashion

Hausburg, T., & Herrmann, Z. (2023, May 1). *Empowering students to be part of climate-change solutions* (Vol. 80, No. 8). ascd.org. https://www.ascd.org/el/articles/empowering-students-to-be-part-of-climate-change-solutions

Kahne, J., & Bowyer, B. (2017). Educating for democracy in a partisan age: Confronting the challenges of motivated reasoning and misinformation. *American Educational Research Journal, 54*(1), 3–34. https://doi.org/10.3102/0002831216679817

Keep America Beautiful litter study ignores corporate blame for plastic pollution—greenpeace USA. (2021, May 19). Greenpeace USA. https://www.greenpeace.org/usa/news/keep-america-beautiful-litter-study-ignores-corporate-blame-for-plastic-pollution/

Kolbert, E. (2017, February). Why facts don't change our minds. *The New Yorker*.

Leverage Point—Tool/Concept/Definition. (n.d.). https://thwink.org/sustain/glossary/LeveragePoint.htm

Liboiron, M. (2018, December 21). How plastic is a function of colonialism. *Teen Vogue*. https://www.teenvogue.com/story/how-plastic-is-a-function-of-colonialism

March, T. (2003, December). The learning power of WebQuests. *ASCD, 61*(4). https://www.ascd.org/el/articles/the-learning-power-of-webquests

McLeod, M., Arp, H. P. H., Tekman, M. B., & Jahnke, A. (2021, July 2). The global threat from plastic pollution. *Science, 373*(6550), 61–65. https://www.science.org/doi/full/10.1126/science.abg5433

Meadows, D. (2008). *Thinking in systems: A primer*. Chelsea Green Publishing.

Monbiot, G. (2021, October 30). Capitalism is killing our planet–it's time to stop buying into our own destruction. *The Guardian*.

https://www.theguardian.com/environment/2021/oct/30/capitalism-is-killing-the-planet-its-time-to-stop-buying-into-our-own-destruction

O'Dell, R., & Marchetti, A. (2021). *A teacher's guide to mentor texts*. Heinemann.

Peer, E., & Gamliel, E. (2013, January). Heuristics and biases in judicial decisions. *Court Review: The Journal of the American Judges Association, 49*, 114–118.

Rosiek, J. (2003). Emotional scaffolding: An exploration of the teacher knowledge at the intersection of student emotion and the subject matter. *Journal of Teacher Education, 54*(5), 399–412.

Sedaghat, L. (2018, April 4). 7 things you didn't know about plastic. *National Geographic*. https://blog.nationalgeographic.org/2018/04/04/7-things-you-didnt-know-about-plastic-and-recycling/

Senge, P. M. (2006). *The fifth discipline: The art & practice of the learning organization*. Currency. (Original work published 1990)

Soares, M. A. (2020, January). Waking up to orwellian spaces: Conscious students and dystopian texts. *English Journal, 109*(3), 74–80.

Stroh, D. P. (2015). *Systems thinking for social change*. Chelsea Green Publishing.

Systems Thinking and the Cobra Effect—Our World. (n.d.). https://ourworld.unu.edu/en/systems-thinking-and-the-cobra-effect

Walsh, J. A., & Sattes, B. D. (2015). *Questioning for classroom discussion: Purposeful speaking, engaged listening, deep thinking*. ASCD.

Weissbourd, R., Manning, G., & Torres, E. (2023, April). Teaching students to talk across political difference. *ASCD, 80*(7). https://www.ascd.org/el/articles/teaching-students-to-talk-across-political-difference

Wiek, A., Withycombe, L., & Redman, C. L. (2011). Key competencies in sustainability: A reference framework for academic program development. *Sustainability Science, 6*, 203–218. https://doi.org/10.1007/s11625-011-0132-6

Wiel, Z. (2021). *The world becomes what we teach: Educating a generation of solutionaries*. Lantern Publishing & Media. (Original work published 2016)

Wolfe, J. (2019, February). Emotional scaffolding: Creating safe spaces for voice. *California English, 24*(3), 18–20.

3
Climate Stewards as Scenario Developers

To address the drought challenges exacerbated by climate change, urban planners continually assess and analyze water resource management needs through the lens of one of our storytelling stances: systemic responsiveness. Anticipating and planning for water network needs is one way urban planners improve the odds of communities being resilient in the future. By studying water resource challenges and encouraging students to predict how challenges will evolve over time, we foster the type of cognitive flexibility needed to both problem-find and problem-solve. This chapter offers an infrastructure scenario for which students must develop solutions. The activities are designed to heighten student awareness of the root causes of water scarcity and deepen understanding of how water flows to our faucet taps.

Paolo Bacigalupi's story "The Tamarisk Hunter" introduces us to Lolo, a water bounty hunter, who makes a living searching for Tamarisk trees in the year 2030. Tamarisks are one of the few remaining sources of water in the perpetually drought-stricken Southwestern desert, as water from the Colorado River is diverted to the wealthy elite in California. Though the story depicts a futuristic society stricken by the challenge of trying to meet needs when water scarcity has changed everything, it did not seem like a far-off possibility.

In literature classes, we typically discuss setting as something that mirrors a character's emotions or helps to establish the mood of a story. However, Bacigalupi's story foregrounds the environment as the principal element. Reading the story is a rare opportunity to consider what it means to share water resources with intergenerational awareness. As we read, we thought about these questions:

> How does the story orient us to understand human accountability to the environment?
> How do environmental stories help us understand accountability to each other?

The story's bureaucratic decision-makers appear ruthlessly calculating: by considering "how many cities, how many people they could evaporate at a time without causing too much unrest," they clearly were shutting off water access in a strategic, gradual manner so as to forestall protests. My students noticed that the word "evaporate" had a double meaning—it both suggested the disappearance of water-stressed cities while reminding us how water liquid can change into invisible vapor.

We can encourage students to look for analogous real-life situations to better understand that the conflicts surrounding sharing and protecting water resources in "The Tamarisk Hunter" are not merely fictional and far off. By inviting students to don their "urban planning" hats, we can ponder what actions need to be taken to reduce dependency on imported water.

Using the VUCA Tool

The acronym VUCA represents four terms: Volatility, Uncertainty, Complexity, Ambiguity. First described in 1985 by economists and university professors Warren Bennis and Burt Nanus, the terms represent the type of challenges leaders face. Volatility refers to the unpredictable nature of change, making it challenging for leaders to prepare for these shifts due to the intensity of fluctuation over time. Uncertainty describes the unpredictability of events. Complexity refers to the number of influencing

TABLE 3.1 Applying the VUCA Tool

What is volatile?	How will extreme weather events increase the risk of waterway flooding?
What is uncertain?	Will drought conditions in the Southwest increase resentment over exported water?
What is complex?	How will competing claims on water resources be resolved?
What is ambiguous?	Is the way we allocate water to agriculture sustainable?

factors and their interactions, making cause-and-effect relationships unclear. Ambiguity refers to the lack of clarity and contradictory, mixed messages that vex our ability to make decisions. Analyzing environmental challenges through VUCA gives students valuable practice in sifting through a high volume of data, much like urban planners do.

The VUCA tool can help students articulate why water management is challenging, yielding a set of questions that can drive their investigation and scenario development.

Reading a story like "The Tamarisk Hunter" brings the issue of water-sharing agreements into focus. The Colorado River is a critical resource in the West because seven basin states (Arizona, California, Colorado, Nevada, New Mexico, Utah, and Wyoming) rely on it for water supply. For more than two decades, climate change has ravaged the Colorado River. The need for cutbacks to out-of-state use and the need to use less water in order to match the declining supply are inescapable realities (Runyon, 2024). To help students consider local alternatives to imported water, we can start by helping them understand the importance of healthy watersheds.

The Journey of Water

> **Box 3.1: Key Term**
>
> Watershed: the land area that "sheds" water to a drainage system or river. A watershed's headwater begins at the mountains and foothills, flows across the valley floor and eventually into a body of water (lakes and ocean).

Students may not know that every land is part of some watershed. Watersheds supply us with water by feeding underground aquifers or channeling water into rivers and other waterways. To help students visualize the journey of water, invite them to create a watershed model illustrating the factors that contribute to harmful runoff.

Activity: Modeling a Watershed

Materials Needed:

- Spray bottles, filled with water
- Two plastic tablecloths
- Crumpled newspaper or paper scraps to create "landscape"
- Small pieces of sponge (representing wetlands)
- Dish soup (representing detergents or cleansers)
- Cocoa powder (representing dirt)
- Honey (representing oil from cars)
- Food coloring (representing fertilizer)

Directions: Lay one of the tablecloths down on a surface. Place the "landscape" items atop the tablecloth, and over the "landscape" with the second tablecloth. The items must be centrally placed on the bottom tablecloth so a runoff effect can be created. Ask students to imagine this is the topography of their community, and point to where landforms and landmarks could be identified. Place the sponge pieces in low spots. Choose a few students to "make rain" by spraying the middle of the tablecloth with the spray bottles.

As the water begins to run down the hills, ask students to imagine which local bodies of water (streams, rivers, wetlands, lakes) will collect it based on how gravity and topography move it to a drain area.

Invite students to add two drops of food coloring ("fertilizer") and spray water, creating "rain."

Invite students to add cocoa powder ("dirt") to one section of the tablecloth, which is loose due to tree removal and will travel with runoff water.

Invite students to drizzle honey ("oil") and dish soup ("cleansers"). Oil leaking from a parked car or cleansers creating soapy water can flow with rainwater into a storm drain.

Then, invite students to spray water and create "rain" again, watching as the water and added substances combine and pocket in different areas, moving in a downward direction.

Though we know gravity moves water from higher to lower areas, the community model can give a sense of some of the obstacles water encounters that interrupt its flow. Pollutants and litter can join water easily, especially in areas where waterways are created through channelization.

Box 3.2: Key Terms

Channelization: the process of engineering waterways to provide for flood control and improved drainage

Point Source Pollution: pollution that is discharged from a single, identifiable source such as pipes, factories, or ships

Nonpoint Source Pollution: pollution that is caused by rainfall moving over the ground as runoff picks up pollutants and deposits them into rivers and other bodies of water.

Activity Reflection: Assessing water pollution through a VUCA lens, describe what makes the problem complex. Identify multiple factors that contribute to this problem.

To help students who need more guidance, ask them to consider:

What are different types of waste collection available in your area?
What kind of car maintenance is needed to reduce oil leaks?
What role can mulch and retaining walls play in solving problems with soil erosion?
What alternatives exist to the use of plant fertilizers?

Scenario Developers and Stormwater

The watershed model offers a great visual for impediments to drainage. When water does not encounter pervious land area, it forms urban runoff. If we identify better ways to capture stormwater locally, we can reduce dependence on imported water.

> **Box 3.3: Key Term**
>
> Stormwater runoff: generated from rain and snowmelt that flows over land or impervious surfaces, such as paved streets, parking lots, and building rooftops.

To help students practice their scenario development skills, invite them to write a letter proposing a plan to better capture stormwater. What would it mean for your city to replace aging water infrastructure with green infrastructure? These brainstorming steps will assist your students in surveying places in their community.

- Study permeable surfaces versus impermeable surfaces: Ideally, where would asphalt or concrete surfaces be replaced with planting beds, mulched beds, gravel, or permeable pavers? Look for surfaces that create an intense heat island effect.
- Consider flood intensity due to extreme weather events: Where do low-lying areas and areas with insufficient drainage systems exist in your community?
- Research health impacts related to water: How does the formation of stagnant water pose health risks?
- Imagine neighborhood revitalization possibilities: How could community gardens designed with permeable surfaces and walkable paths increase a sense of collective well-being?

Task: Identify the best way to communicate a message about protecting local water resources (speech, letter, etc.). Direct your

TABLE 3.2 Letter Checklist

Letter Rubric Checklist
Writer offers details for converting a gray infrastructure system to a green infrastructure system.
Writer stresses the importance of permeable surfaces.
Writer communicates the connection between sustainable stormwater management and climate change mitigation and adaptation.
Writer suggests the benefits of a new infrastructure system for neighborhood revitalization.

message to the audience who is in a position to effect change about this matter.

In *Writing Rhetorically: Fostering Responsive Thinkers and Communicators*, Jennifer Fletcher offers a transferable process for rhetorical problem-solving:

> What's the problem? (exigence)
> What do you want to do about it? (purpose)
> Who has the power to make this change? (audience)
> What's the best way to reach this audience? (genre)
> Why is now the right time to act? (*kairos*)
> (2021, p. 59)

Using Fletcher's set of questions, we can better tailor and direct our letter to the appropriate addressee.

Directions: Write a letter to your city's Director of Public Works, proposing a plan to better capture stormwater by replacing aging infrastructure with green infrastructure.

Storytelling Beyond Either/Or

It can be easy to think about water scarcity in California in a monolithic way: imagining that every region faces the same challenges when it comes to implementing equitable water management practices. One of the most contentious environmental issues is the agricultural use of water. According to the California Department of Food and Agriculture, over a third of the country's vegetables and over three-quarters of the country's fruits and nuts are grown in California. California's Central Valley is the heart of agricultural production, where rising temperatures, constrained water resources, and increased disease pressure are causing many growers to consider growing more resilient varieties of crops. When we "zoom out" and think about the impacts of climate change, water journalists such as the *Los Angeles Times*' Ian James and Hayley Smith paint an especially bleak picture for agriculture. Distilling insights from the most recent U.S. Climate Report, they state:

> Extreme heat, drought and water scarcity are projected to have severe or worsening effects on agriculture in California, reducing crop yields, harming orchards, increasing heat stress for livestock, and shifting zones where crops can be grown–all of which carry major economic costs.
>
> (2023)

This leads me to wonder: are farmers unfairly blamed for taking too much water?

In many cases, groundwater depletion is the cause of issues related to water scarcity. Many farming communities are grappling with chronic groundwater depletion, as seen in the current conflict in the Cuyama Valley. The legal ramifications of these conflicts are challenging the state's efforts to implement pumping limits and other requirements of a law put into effect in 2014, the Sustainable Groundwater Management Act. The Cuyama Valley is one of 21 groundwater basins that the state has deemed "critically overdrafted," a circumstance that threatens the ability of local agencies to develop plans to halt over pumping and stabilize groundwater levels by 2040, as required legally. In this region, "achieving these goals is expected to require slashing water use by as much as two-thirds" (James).

Overdrafted water basins have implications for the ability to allocate water equitably, especially when different stakeholders contest the terms of allocation. An intense legal war over water rights has arisen between carrot growers and residents in the Cuyama Valley. Residents have accused Grimmway Farms and

Bolthouse Farms, the valley's biggest water users, of going to court to secure as much water as possible while forcing painful cuts on smaller farms. Due to this hardship, residents have organized carrot boycotts, resulting in both companies dropping out of the lawsuit and filing requests to have their names removed as plaintiffs. The outrage residents felt was in large part driven by the fact that together, the "companies used nearly three times the annual water use of the city of Santa Barbara" (James). Small farmers share anger and resentment when they discuss how they have done their part and switched to less water-intensive crops such as olive trees, while witnessing the carrot growers expand and drill more wells. There is a rampant sense that these large carrot companies are exploiting the area for economic profit, uncaring of the needs of community residents and straying from the idea of eating what is seasonal (O'Connor, 2023).

Discussion Prompt: Dig Deeper

Small farms are feeling the stakes of ongoing water scarcity differently than large farms in the Cuyama Valley. How can we tell the story of agricultural water use with more nuance when naming environmental culprits?

Problem-solving With a Different Sense of Time

We can help students gain a foothold in their research investigation by modeling how to situate the issue in context. The historic use of water and existing water treaties play a huge role in how the state of California legally determines water ownership and rights. When asking the question—who makes decisions about how water is allocated and shared?—educators can indicate that legal claims going back decades frequently trump the concerns of contemporary water councils and water justice advocates.

Educators can encourage investigation and collective brainstorming by asking groups of students to zoom in on a regional or local water scarcity problem that addresses the questions: How can we manage our water supply equitably and avoid the problems with water privatization seen in this story and already happening in our home state? What water efficiency models can be replicated in a wide-scale manner? Prior to beginning their investigations, educators can model with a whole-class case study, such as with the Cuyama Valley carrot boycotts.

To engage students in taking action on a topic that is both relevant and a looming emergency, we can pair newspaper headlines with fictional reads. Neal and Jarrod Shusterman's novel *Dry* is set in the South Coast region of Southern California, which is depicted as relying too much on imported water. The expectation that water from the Colorado River would flow west in perpetuity is identified as one of the major reasons for the apocalyptic conditions besetting the novel's characters. By treating the river as a "lifeline," one character says in disgust, Californians made themselves vulnerable. Reading a fictional account of the day the water supply runs out in Southern California—and all the chaos that ensues—makes us uncomfortably aware of how little we know about the water we enjoy: where it comes from, how it's managed, and what we are doing to mitigate prolonged drought conditions in our state.

When pairing fiction and non-fiction, we can ask students to weigh the roles of hindsight and foresight in environmental problem-solving. Because dystopian storytelling frequently shows us the link between reckless human behavior and environmental degradation, we can develop foresight about the best way to augment local water supplies, as discussed in the article. Alyssa Morrow, the main character of *Dry*, speaks directly to the role of hindsight:

> In hindsight, we should have come straight here the moment the taps were turned off. But when something drastic happens, there's a lag time. It's not quite denial, and not quite shock, but more like a mental free fall. You're spending so much time wrapping your head around a problem, you don't realize what you need to do until the window to do it has closed.
>
> p. 12

The dwindling window of opportunity becomes clear in retrospect, which is why it is so important to not only consider the symptoms of climate change but its root cause: the high amount of atmospheric carbon dioxide causing warmer temperatures, due to the burning of fossil fuels. The novel shows the impossibility of FEMA addressing the needs of hurricane victims and the needs of the water shutoff victims simultaneously. Because climate change makes it likely that we'll be navigating multiple emergencies at once, we need to develop foresight to build our collective adaptive capacity.

Protecting Water

Mary Ann Ng is a science teacher at Alhambra High School in Alhambra, California. "When students understand the lesson topic is local, they are more engaged." She experienced this herself when she began to travel again after post-pandemic isolation. She visited the Owens Valley and the Klamath River, places associated with California "water wars," which deepened her understanding of how scientific concepts and social issues are intertwined.

I asked Mary Ann about opportunities to bring tribal perspectives into a science classroom. She describes the Introduction to Phenomena stage occurring at the beginning of a unit as a key opportunity. "Invite students to look at the Fallon Paiute Shoshone Tribe Seal. Then pair it with pictures of drought-stricken land in the tribal area." This phenomenon pairing can invite students to think about which perspectives are centered in environmental stewardship and which perspectives have been left out historically. "We're not propelled to ask questions or take action unless we're pushed out of our comfort zone. Teachers have an opportunity. We can't just hand out textbooks."

Like Mary Ann, I bring in pictures, videos, and other texts to encourage student speculation. Through juxtaposing a picture book about the Water Protectors at Standing Rock with other texts, I hoped to broaden our vision of what constitutes source relevance.

The first picture book I shared with my high school students was Carole Lindstrom's *We Are Water Protectors*. The line "The river's rhythm runs through my veins" signals a way of thinking about water that suggests kinship. My students were intrigued by Michaela Goade's book note explaining how she honored Lindstrom's Ojibwe culture through key visual details: the traditional ribbon skirt worn by the protagonist as she rallies her people, the animal imagery reflecting Anishinaabe/Ojibwe clan symbols, and the repeated floral designs inspired by traditional Anishinaabe woodland floral motifs. Goade, an enrolled member of the Tlingit and Haida Indian Tribes of Alaska, is the first Indigenous illustrator to win the Caldecott.

Throughout the book, Goade's use of negative painting, watercolor, colored pencil, and gouache beautifully compose images embodying the Lakota phrase "Mni wiconi" ("Water is life"). By using *We Are Water Protectors* as our anchor text for a research case study on the Standing Rock Sioux Tribe's stand against the Dakota Access Pipeline, I aimed to see what students would notice about media representations of Indigenous resistance: what was emphasized, what was left out, what was conflated.

One important noticing that I heard echoed across groups regarded the name given to the people gathered at Standing Rock. Most of my students had never heard the phrase "Water Protectors" before. They had noticed Indigenous writers would use the phrase, but nearly every other news source fell back on referring to the camp activists as "protestors." The difference seemed innocuous, at first, but one student drew our attention to a point made by Roxanne Dunbar-Ortiz in *An Indigenous Peoples' History of the United States for Young People*: "The pipeline company, law enforcement, and the news media usually called them 'protestors,' which emphasized objection to the pipeline rather than the goal of saving the region's water supply" (2019, p. 212). This noticing prompted an interesting discussion of how the names we give ourselves can show what we value. The term "protectors," as seen so vividly in Lindstrom's book, reveals a commitment to defending the sacred and experiencing kinship with all living things.

What began as a question about why the Standing Rock Sioux Tribe was fighting DAPL evolved into a larger question about how a coalition of Water Protectors could help others reconceive their relationship with the Earth's water. What mindset shift needed to take place? By addressing harmful mindset assumptions about water's availability as a commodity, my students were tackling the root cause of so much environmental exploitation.

> ### A Peek Into Their Practice
>
> Generation Earth is a Los Angeles-based environmental education program that helps students explore water, waste, and forestry topics. Project experts from the Generation Earth team are willing to come to a school site and work with teachers to introduce these topics with engaging lesson ideas. When I reached out for assistance with lessons focused on water resilience, Hillary Michelle visited my 11th-grade English class and shared ideas for helping students understand the flow of water. Hillary is grateful to Generation Earth for helping her engage in collective action: "Having the opportunity to partner with schools and engage with students on various environmental topics has been amazing. I enjoy watching their faces light up when they learn new information and see that they can have an impact in their communities and on the world."
>
> Hillary offered these steps for conducting a water audit on our own school campus. Through creating a map indicating where specific water-related elements are available on campus, students are set up to investigate the direction water takes and identify any areas of concern.
>
> **Materials for student groups:**
> - A large sheet of paper to create a handmade map
> - Colored pencils/markers

Directions:

Students gather data by color coding specific water-related elements.

- Use Green to show trees and places where water can get into the ground (bare dirt, garden, grass)
- Use a Black dot to show existing trees on your map
- Use Blue to show sources of water (faucets, sprinklers, hoses)
- Use Purple to show places where water travels (gutters, downspout, drain)
- Use a Red X to show trash and other things (food, oil) that could be harmful to water

Afterwards, use arrows to show the direction water takes and identify any areas of concern. Be sure to take note of where water pools during a rainstorm, where you found a lot of trash, and possible sources for where the trash came from.

Give student groups an opportunity to interpret the data and identify next steps for addressing water pollution on campus. Decide together who is an appropriate person to contact regarding issues such as clogged drains, flooded areas during water runoff, and empty tree wells.

Guiding Question: How can a water audit help us identify problems on campus?

Learning Standards

Next Generation Science Standards—Earth Systems:

HS-ESS2–5 Plan and conduct an investigation of the properties of water and its effects on Earth materials and surface processes.

CCSS.ELA-Literacy.SL.11–12.1.c:

Propel conversations by posing and responding to questions that probe reasoning and evidence; ensure a hearing for a full range of positions on a topic or issue; clarify, verify, or challenge ideas and conclusions; and promote divergent and creative perspectives.

Connections to Crosscutting Concepts:

Scale, Proportion and Quality—developing a model to represent patterns in the natural world
Cause and Effect—determining the cause and effect of high water usage
Stability and Change—identifying areas where change is needed to improve water management

Conclusion

The ability to transfer learning to new tasks is key to being the type of adaptive thinker needed to problem-solve. By viewing water challenges through a VUCA lens, students gain valuable practice in assessing planning challenges. To empower learners to take action on water scarcity, it is important for educators to find entry points for students to sufficiently consider the stakes of the issue in light of climate change.

References

Bacigalupi, P. (2006). The tamarisk hunter. *High Country News*. https://www.hcn.org/issues/issue-325/tamarisk-hunter-bacigalupi/
Bennis, W., & Nanus, B. (1985). *Leaders*. Collins Business.

Dunbar-Ortiz, R. (2019). *An Indigenous peoples' history of the United States for young people*. Beacon Press.

Fletcher, J. (2021). *Writing rhetorically: Fostering responsive thinkers and communicators*. Stenhouse.

Generation Earth. (2023). *Toolkit: Water pollution prevention*. Los Angeles County Public Works.

James, I. (2023, November 18). Soaked in controversy: Big carrot guzzles water in large volumes. Residents, farmers counter with a boycott. *Los Angeles Times*.

Lindstrom, C., Goade, M., & Fott, G. (2021). *We are water protectors*. Scholastic.

O'Connor, T. (2023, October 12). Basin battle: Cuyama Valley's small farmers, landowners face off against corporations over groundwater. *New Times San Luis Obispo*.

Runyon, L. (2024). *What to watch on the Colorado river in 2024*. The Water Desk. https://waterdesk.org/2024/01/what-to-watch-on-the-colorado-river-in-2024/

Shusterman, N., & Shusterman, J. (2018). *Dry*. Simon & Schuster.

Smith, H., & James, I. (2023, November 24). 'Every bit matters': Six key takeaways from the latest U.S. climate report. *Los Angeles Times*.

4

Climate Stewards as Environmental Justice Activists

Acknowledging historic and ongoing disparities in how communities are protected from environmental hazards is a critical aspect of cultivating climate stewards. An approach that tackles climate change challenges without considering the disparate impacts to communities ignores the health threats compounded by inequities embedded systemically through racist city planning and housing practices, inadequate protection due to aging infrastructure, and siting of toxic wastes in vulnerable communities. If we can help students broaden their conception of environmental protection, then they will be better prepared to consider socioeconomic factors as they problem-solve, folding a more nuanced awareness into their communication they deliver as climate stewards. To help educators demystify the practice of environmental justice, this chapter offers a guide for cultivating these skills:

- Ability to identify factors that contribute to different risk burdens
- Ability to communicate why environmental hazards are public health emergencies
- Ability to identify mechanisms for accountability
- Ability to advocate for prevention of adverse health impacts

Without being mindful of these skills, students might default to reductive thinking or superficial glossing of what is at issue when thinking about the causes at the root of environmental injustices. Consider the case of the Flint Water Crisis. The state of Michigan in April of 2014 switched the city of Flint from its prior water supply of Lake Huron, provided by Detroit, to the Flint River. No corrosion inhibitors were applied to the water, allowing lead from the aging pipe infrastructure to leach into the water supply used for drinking and bathing (Clearfield & Tilcsik, 2018). Around 100,000 residents were exposed to elevated lead levels, even as government officials knew about the water contamination. Nearly 30,000 children were exposed to a neurotoxin known to have detrimental effects on children's developing brains and nervous systems (Green, 2019). Children are particularly vulnerable to the long-term effects of lead poisoning, which can include a reduction in intellectual functioning and an increased risk of Alzheimer's disease.

When applying an environmental justice lens, the focus goes to the source: understanding the experience of community residents directly affected by an environmental hazard, in their own words. "Apparently the city of Flint's water quality is not good enough to be used in an industrial process but good enough to [be] used and consumed by humans," wrote D'Andre Jackson in a letter to the *Flint Journal*. As soon as it became apparent that Flint River water rusted engine parts at the General Motors plants in Flint, city officials agreed that the engine factory could reconnect with Detroit-treated Lake Huron water. The failure of national news outlets to take seriously the concerns of Flint community members occurred in marked contrast to local media voices, who noted the fortitude and fight of residents advocating for themselves and access to healthy water. When analyzing coverage of the crisis, Derrick Z. Jackson identifies multiple factors playing a role in this failure: "a lack of newsroom diversity, a history of national media paying little attention to environmental justice in communities of color, and the tendency to act only after harm has been verified by doctors and scientists" (Jackson). Nicky Sheats, a co-founder of the New Jersey Environmental Justice Alliance, offers an illuminating criticism of the coverage: "Whether it's

Flint or Katrina, these stories read like reporters discovered disparities for the first time" (Jackson). By overlooking the concerns of Flint residents and covering the health emergency in a manner that merely skimmed the surface, some state and national media neglected to connect the dots between the lack of urgency and environmental racism, only sounding the alarm when medical experts confirmed the claims of residents.

What Is Environmental Justice and Environmental Racism?

The environmental justice movement, emerging in the 1970s and 1980s, investigates and protests the racialized and uneven exposure to urban environmental health hazards and risks. In 1987, the Commission for Racial Justice's landmark study, *Toxic Wastes and Race*, found race to be the single most important factor (i.e., more important than income, the percentage of people who own their homes, and property values) in the location of abandoned toxic waste sites. Environmental racism can be defined as

> the institutional rules, regulations, policies, or government/corporate decisions that deliberately target certain communities for locally undesirable land uses and lax enforcement of zoning and environmental laws, resulting in communities being disproportionately exposed to toxic and hazardous waste based on race.
> (greenaction.org)

In his groundbreaking work, *Dumping in Dixie: Race, Class, and Environmental Quality*, Robert D. Bullard points out that the dominant environmental protection paradigm—emphasizing probability of fatality as a model for decision-making—ignores environmental stressors that produce dangerous, non-deadly health conditions. These include developmental, reproductive, respiratory, neurotic, and psychological effects. Bullard raises questions about the inequitable outcomes of environmental practices that the mission of the federal Environmental Protection Agency was not designed to address:

> Who is most affected?
> Why are they affected?
> Who created the problem?
> What can be done to remedy the problem?
> How can the problem be prevented?
> (Bullard, 2000, p. 116)

These questions can serve as an important touchstone in the classroom as students make their initial forays into research investigation. They elicit answers based on contextualizing an environmental injustice, not merely skimming the issue through basic journalistic information gathering. As students advance through stages of an investigation task, educators should encourage them to revisit these questions as they identify mechanisms for keeping people healthy and protected in frontline communities.

Communities made vulnerable based on proximity to environmental hazards live in areas called "sacrifice zones": hotspots that expose residents to dangerous chemicals and other environmental threats, where community health is surrendered in the name of profit. Raising the question of who pays for and who benefits from the current environmental and industrial policies is a key aspect of analyzing how environmental racism plays out systematically. "Flint was under emergency management when critical decisions about the city's water supply and water treatment protocols were made, and emergency managers were resistant to public concerns about the safety of the city's drinking water" (Hughes et al., 2021). The austerity policies of emergency managers made drinking water safety and infrastructure vulnerable, demonstrating that state intervention addressing financial distress can dangerously prioritize short-term budget balancing over the health concerns of community residents.

The demand for equity in how we respond to environmental hazards is an essential aspect of practicing environmental justice. Looking at case studies like the Flint crisis will help to orient your students as they begin to discern some common denominators existing in case examples of environmental injustice. Giving students practice in identifying indicators of environmental health hazards is an important step in supporting their efforts in researching case studies on their own.

Activity: Broadening Our Understanding of Environmental Protections

What do educators need to know in order to help students look at environmental hazards from an environmental justice lens? The first activity for this chapter guides you to reflect on three instances of water contamination: occurring in Flint, Michigan; Denmark, South Carolina; and Newark, New Jersey. Looking at a text set with students at the beginning of a research unit offers a meaningful entry point to online investigation. Before researching on their own, students need to have habits of inquiry modeled for them (Coombs & Bellingham, 2015).

Activity 1 Text Set

Text 4.1
   Bates, J. (2019, August 27). Newark officials providing bottled water to 15,000 homes over lead contamination concerns. Here's what your need to know about the city's water crisis. *Time.* https://time.com/5653115/newark-water-crisis/

Text 4.2
   Ganim, S. (2018, November 28). For 10 years, a chemical not EPA approved was in their drinking water. *CNN.* https://www.cnn.com/2018/11/11/health/denmark-sc-water-chemical-not-epa-approved/index.html

Text 4.3
   Vaughen, K. (2023, March 23). Michigan still dealing with fallout from Flint water crisis 9 years later; plus new waterworries. *CBSNews.* https://www.cbsnews.com/detroit/news/michigan-still-dealing-with-fallout-from-flint-water-crisis-9-years-later/

TABLE 4.1 Factors That Contribute to Different Risk Burdens

Types of Indicators	Environmental Health Hazards
Exposure indicators	• Air quality: Ozone • Air quality: PM2.5 • Children's Lead Risk from Housing • Diesel Particulate Matter • Drinking Water Contaminants • Pesticide Use • Toxic Releases from Facilities • Traffic Impacts
Environmental effects indicators	• Cleanup Sites • Groundwater Threats • Hazardous Waste Generators and Facilities • Impaired Water Bodies • Solid Waste Sites and Facilities
Sensitive population indicators	• Asthma • Cardiovascular Disease • Low Birth Weight Infants
Socioeconomic factor indicators	• Educational Attainment • Housing Burden • Linguistic Isolation • Poverty • Unemployment

Source: Adapted from CalEnviroScreen

After reading the three articles, direct students to study Table 4.1 and consider the various indicators of environmental health hazards. Then, reflect on this question together:

> What does this set of texts help you understand about how residents in these three communities were exposed to drinking water contaminants?

Activity: Comparing and Contrasting Instances of Environmental Injustice

Invite students to fill out the organizer in Table 4.2 based on their reading.

A Compare/Contrast Matrix Graphic Organizer is a useful tool for looking at more than two items and helps students create

Climate Stewards as Environmental Justice Activists ◆ 111

TABLE 4.2 Compare/Contrast Matrix Graphic Organizer

Questions	Newark, New Jersey	Denmark, South Carolina	Flint, Michigan
Who is most affected?	Children are most vulnerable. Residents must rely on water filtration. Advised to use bottled water for cooking, drinking, and preparing baby formula	Children are most vulnerable. Residents have been complaining about skin rashes and kidney-related issues	Children are most vulnerable. Nine years after the water crisis, many residents still avoid tap water. They rely on bottled water to brush their teeth and feed their pets
Why are they affected?	Toxic lead levels in water as a result of pipe corrosion	HaloSan, a non-EPA approved chemical, was used to treat water	Trust has been broken and the lead service lines have not been replaced entirely
Who created the problem?	Aging pipe infrastructure that needed to be replaced through federal funding	South Carolina's Department of Health and Environmental Control approved its use	The former governor, some members of his office, some state emergency managers, some members of the Michigan Department of Health and Human Services
What can be done to remedy the problem?	Improved corrosion control	Additional federal grants to replace aging pipes	Complete replacement of lead service lines, as required of a 2017 settlement
How can the problem be prevented?	Replacing lead pipes with copper pipes	Greater transparency from government officials and less reliance on a quick fix	Anticipating problems with water quality that are exacerbated by weather events; completing pipe overhaul

a record of the salient details they pulled from their readings based on Bullard's line of questioning. The questions offer clear criteria that keep students focused on relevant information, leading them to weigh ideas about what proactive prevention should look like. After completing the organizer, educators can model making connections between the three examples of environmental injustice, noting that all examples relate to the same exposure indicator: drinking water contaminants. The article reporting on the Denmark water crisis brings up an important socioeconomic factor indicator when mentioning lower population numbers means decreasing revenue and rising water costs. The article reporting on lingering concerns and mistrust in Flint brings up holistic concerns concerning water quality, given the detection of forever chemicals in the Great Lakes region. This activity can demonstrate to students the usefulness of recording comparisons and contrasts when weighing and synthesizing information from multiple sources.

Featured Mentor Text: *Mayah's Lot*

In Charlie La Greca and Rebecca Bratspies's environmental justice comic, *Mayah's Lot*, the image of the aspen seed is prominent. The titular character intends to plant an aspen seed in a garden she secretly tends on a vacant lot, just before finding out a corporation's plan to transform the lot into an industrial toxic storage waste facility. The seed growth imagery symbolizes how the work of environmental justice can be achieved: Mayah's voice is joined by others in her community as they feel increasingly empowered to influence decision-making conversations affecting their collective health. As Mayah's neighbor, Mr. Tatsumi, explains to her in a foreshadowing image, "One Aspen seed can create an entire forest." Beautifully illustrated by La Greca, the comic culminates in a page-length panel of Mayah's community, imagined through joyful scenes of her neighborhood as a strong root system extending out in an infrastructure of support.

The "seed work" presented in *Mayah's Lot* can cast light upon the fact that many public health emergencies exist as a result of environmental racism. Reading this environmental justice comic can help students grasp this connection: people existing

TABLE 4.3 *Mayah's Lot* as a Model for Environmental Justice Activism

Skill Target	Type of Comic Shot	Mayah's Lot
Ability to identify factors that contribute to different risk burdens	Establishing shot: a panel that often comes at the beginning of a scene to convey basic facts about setting, such as the characters' location and when and where the scene will be taking place	Panel depicts a lot on a street corner full of litter and a small, carefully tended garden plot in Forestville, New York. A sign near the fence announces, "The Future Home of Green Solutions"
Ability to communicate why environmental hazards are public health emergencies	Close-up shot: a panel that shows an image in a large view, often focusing on a character's face	Panel depicts Mayah's shocked expression at the sight of her destroyed garden and the blueprint plans for toxic waste storage
Ability to identify mechanisms for accountability	Long shot: a panel that depicts an object fully, from top to bottom, such as a character's full body, from head to toe	Panel depicts Mayah's lone figure as she arrives at her first community meeting
Ability to advocate for prevention of adverse health impacts	Medium shot: a panel that shows action or dialogue, typically among two to three people, or a person interacting with their surroundings	Panel depicts three members on the neighborhood council, with one council member opening the floor discussion, as seen in the speech bubble

in frontline minority communities disproportionately experience the burden of environmental hazards. Thinking about the sequential organization of the comic as it presents Mayah's journey into environmental justice activism offers a useful model for our students as they research health threats to communities and identify mechanisms for protecting residents.

Activity: Reflecting on the Mentor Text

How did panels relying on establishing shots convey distinct settings in the comic? What differences could you note between panels depicting scenes in Mayah's neighborhood and scenes associated with the company outsiders?

How do close-up shots give readers a sense of the story's emotional stakes? What do close-up shots allow readers to notice?

What important information did panel captions communicate? What is ironic about the name of the company wishing to store toxic waste in the Forestville neighborhood?

How did attending the community meeting function as a key shift in Mayah's character development?

Why were medium shots an effective choice for depicting mobilized members of Mayah's community?

In discussing the genesis of the comic book, author Rebecca Bratspies describes how her conversations with students revealed their understanding of the environment as a "thing" that exists elsewhere apart from where they lived: "For these students, 'the environment' meant only green places, pristine landscapes with animals and trees. As a result, their conception of environmental protection was not about themselves or their communities" (p. 507). Through studying the comic panels juxtaposing scenes of mobilized neighborhood advocates and scenes of boardroom market share discussion, students can learn about the motivations of vying stakeholders. Through studying its panels showing Mayah in close-up shots or surrounded by speech bubbles, they learn a different model of the hero's journey: one not predicated on "going at it alone" but on building a coalition of like-minded community members eager to protect themselves. The emotional stakes of fighting back becomes clearer and teaches students about a broader, intersectional environmentalism where social justice and environmental justice are not siloed.

The environmental justice movement "shifts the focus of policies from the degradation of 'nature' to address the simultaneous degradation of communities" (Méndez, 2020, p. 25). The mainstream environmental movement many people are familiar with does not usually educate about the type of cruelty represented by the character Lulu in *Mayah's Lot*—a character whose name also functions as an environmental acronym (Locally Undesirable

Land Use, or "LULU"). Lulu wishes to site "locally undesirable" toxic waste in the Forestville lot that Mayah has transformed into a garden, even striking down a proposal for an alternative site in Watertown where "way too many homeowners" are sure to fight back. Teaching this term can deepen our students' environmental literacy and give them the vocabulary to identify LULUs that exist where they live.

Using the Power of Storytelling To Inform

Comic book storytelling can be a useful tool to understand environmental degradation from an equity lens. This became clear to me during a teaching challenge in my own classroom. As I listened to the results of my students' research investigation on local examples of environmental injustice, I observed that their oral presentations were notable for their listing quality: my students conveyed data point after data point. I heard presentations listing facts about a manufacturing plant that produces pesticides that have been banned in the United States, listing facts about an unchecked hazard waste incinerator, and listing facts about a planned waste-to-energy incinerator in South Central Los Angeles. Each speaker presented an accretion of data points that built the speaker's ethos, but did not do much in terms of sustaining the listener's interest. I wondered how I could give students skill practice in creating a story and help them understand the usefulness of *telling a story* while trying to convey information. They unquestionably had met the rubric task of synthesizing source material to inform their audience, but their presentations failed to be memorable.

My students' presentation style was a result of not knowing how powerful research investigation can be when presented in a narrative structure. In his book, *Houston, We Have a Narrative*, Randy Olson describes how many scientists face a similar challenge:

> The tendency of scientists to present endless piles of facts, unable to find the singular narrative onto which everyone

can focus, has become a reason many important science stories, including that of global warming, fail to resonate with the public.

(p. 42)

Olson identifies the crux of the problem: scientists have important information to convey, but many fail to do it in a compelling manner. If we are truly committed to helping this generation of students become scholars, they must not just be trained in content, but also in ways of communicating with an audience. The fact that communities of color are routinely targeted to host facilities that have harmful environmental impacts is often overlooked. It is a type of erasure that downplays how environmental history intersects with a social history of race and class inequity. We can help our students be credible sources of information about these harmful impacts, while moving beyond a mere focus on data presentation.

Creating Storyboards

Mayah's Lot served as a mentor text for my students. It beautifully models how a narrative can stage both a problem and possible solutions. To help students incorporate storytelling when presenting key information, I gave them opportunities to create storyboards: that is, visual narratives where instances of environmental injustice and solutions can be presented. Storyboarding can support students' metacognitive understanding, aiding their mental rearrangement of key research details as they distill facts from consulted sources (Jacobs & Zmuda, 2023). Telling a story via images can pare down extraneous verbal information, helping the audience understand the storyteller's explanation in a visualizable form that can be a memory tool. *Storyboard That* has a drag-and-drop platform that has been essential in teaching my students how to create sequential narratives about their research. Through using this web-based digital storytelling tool, students can create stories about topics that otherwise can seem abstract and complex. The same opportunities for developing inferential skills exist if encouraging students to storyboard with hand-drawn illustrations. As they fill in their panels with drawings,

TABLE 4.4 Environmental Justice Storyboard Rubric

	4	3	2	1
Produce storyboard that convey reasons for an environmental hazard	Storyboard narrative conveys causes of an environmental hazard due to inequitable burdens.	Storyboard narrative clearly conveys causes of an environmental hazard due to inequitable burdens.	Storyboard narrative conveys at least one cause of an environmental hazard due to inequitable burdens.	Storyboard narrative insufficiently or unclearly conveys a cause of an environmental hazard due to inequitable burdens.
Create a sequence of panels that convey mechanisms for accountability	Sequence of panels specific mechanisms for accountability based on varied, credible sources.	Sequence of panels convey specific mechanisms for accountability based on credible sources.	Sequence of panels convey specific mechanisms for accountability based on one credible source.	Sequence of panels insufficiently convey specific mechanisms for accountability.
Delineate how community members engage in self-advocacy	Varied comic book elements effectively delineate how community members engage in self-advocacy.	Varied comic book elements delineate how community members engage in self-advocacy.	At least two comic book elements delineate how community members engage in self-advocacy.	Storyboard insufficiently delineates how community members engage in self-advocacy.
Communicate solutions based on research	Storyboard narrative clearly conveys solutions to an environmental injustice based on research.	Storyboard narrative conveys solutions to an environmental injustice based on research.	Storyboard narrative conveys at least one solution to an environmental injustice based on research.	Storyboard narrative insufficiently or unclearly conveys a solution to an environmental injustice based on research.
Demonstrate command of comic shots and gutter transitions	Use of varied comics shots and gutter transitions adds to clarity and development of environmental justice narrative.	Use of comics shots and gutter transitions develops environmental justice narrative.	Some errors with use of comic shots and gutter transitions interfere with comprehensibility.	Many errors with use of comic shots and gutter transitions interfere with comprehensibility.

TABLE 4.5 Planning Tool: From Storyboard to Advocacy

Task	Checkpoints
Gathering Sources	• Create Padlet or slide deck of source links • Consult varied and credible sources
Reflecting on Peer Feedback and Annotations	• Make note of patterns in feedback • Prioritize investigation goals and refine search terms
Exploring Data Visualization	• Document findings in your own words • Make connections with "slow violence"
Identifying Policymakers and Decision-makers	• Research policy levers • Assess the problem through different scales
Using A-B-T Storytelling Method	• Shrink research narrative into discrete parts • Practice using the A-B-T structure in oral storytelling
Employing Rhetorical Devices in Direct Appeal	• Study craft moves of relevant mentor texts • Choose rhetorical devices to organize call to action

educators can raise the question: How does the space between the gutters convey meaning just like the spaces between words (Fisher & Frey, 2013)?

Encouraging students to create storyboards in the middle of a research investigation poses multiple benefits for their learning. By previewing the storyboard rubric together in advance, the feedback we give to our students can be aligned with criteria shared ahead of time. The storyboard serves as a formative assessment, assigned mid-research investigation, which allows educators to give feedback that can still be acted upon (William, 2013). Educators can review specific vocabulary related to comic storyboard layouts, point the way to previously unexplored databases, and ponder the feasibility of the environmental justice solutions posed through visuals with students, side by side as "thought partners."

One of the advantages of planning a storyboard before creating a call to action tailored to a specific addressee is students already have a visual representation of their main talking points: now, they need to fill whatever investigation gaps remain. Throughout their investigation, students can keep track of their sources by capturing their hyperlinks on a Padlet or a slide deck.

Creating classroom conditions where students can push each other to refine their research questions depends on them having opportunities to mull provocation to deeper thinking together. For that reason, students circulate and read each other's storyboards, leaving questions and comments on Post-it notes before moving on to a new storyboard. Structured peer-assessment that is focused on improvement rather than grading can be a useful resource for every student in the room, as they have been experiencing trial-and-error methods in research and understand common pitfalls (William, 2013).

Educators can model storyboard annotation by first modeling their own questions and comments aligned with the rubric. To encourage thinking about how students delineated community members engaged in self-advocacy, we can ask: Does the storyboard suggest community agency or community helplessness? This prompt can activate prior knowledge, building on earlier class discussion regarding how residents in Flint, Michigan, Denmark, South Carolina, and Newark, New Jersey fought to protect themselves. To ensure students have adequately researched, we can ask: Does the storyboard communicate risk burdens specific to a place or community? If solutions posed seem hastily conceived and appear too early in the sequence of panels, we can ask: Who else needs to appear in a comic shot for the audience to easily grasp the call to action? After students have read the Post-it annotations made by their peers, give them time to reflect on peer feedback and prioritize new goals at this juncture of their research.

Identifying Mechanisms for Accountability

When students begin to think about mechanisms for accountability, they might assume the government agencies and departments that operate as safeguards always are keeping an eye on public health, safety, and protection. The stories coming out of places like Denmark, South Carolina and Flint, Michigan disprove that assumption. While digging deeper into the research, they will notice the collaborative impetus for mobilizing for environmental justice mostly comes from grassroots organizations who are fighting to make these mechanisms work on behalf of their communities. As students turn from storyboarding and class presentations, they step closer to identifying those policymakers and decision-makers to whom they wish to direct their message of advocacy. Recognizing that people living in frontline communities are experts means much of the research journey involves going back to the source—learning from residents in their own words.

As a result of reading Sharon Lavigne's testimony before the U.S. House Subcommittee on Environment and Climate Change, many students were moved to research Cancer Alley, an 85-mile stretch of land along the Mississippi River between Baton Rouge and New Orleans that is home to over 150 petrochemical plants and refineries. Seven of the ten census tracts with the country's highest cancer risk levels from air pollution are in Cancer Alley. Here, breathing problems, noxious smells, and cancer threats disproportionately impact predominantly Black communities, where the majority of the polluting facilities operate (Greenfield). The testimony of residents such as Lavigne communicates the harmful effects of living near the combined sources of so much industrial pollution. The daughter of civil rights advocates and a lifelong resident of St. James, Lavigne founded RISE St. James, a faith-based grassroots environmental organization and began mobilizing opposition to a newly proposed plastics manufacturing plant for the Chinese chemical company Wanhua. Her actions offer an inspiring model for grassroots environmental justice activism:

- Regularly attended parish council meetings and hearings, asking questions about the proposed plant
- Organized door-to-door visits in neighborhoods that would be most impacted
- Hosted town halls where trusted experts could educate the community
- Produced reports, wrote letters to regional newspapers, and designed newspaper ads in opposition to the project
- Met one on one with council members to persuade them to rescind permits for the plant
- Led marches to raise visibility on the issues
- Built coalitions with local civic organizations and church groups

- Forged links with environmental and climate justice organizations in Louisiana and nationally

(goldmanprize.org)

A year after the plant proposal and increasing public scrutiny, Wanhua canceled the project. In her 2019 congressional testimony, Lavigne vowed to fight a new threat: Formosa Plastics' plan to build a massive ethane cracker complex near her home.

By reflecting on all the actions Lavigne took to protect St. James Parish, my students developed a broader understanding of how calls to action could be directed at actors at different scales:

Local: Neighborhood councils
Regional: Newspaper editors
State: Environmental justice organizations
National: Members of Congress

FIGURE 4.1 A Comic Cell Created via *Storyboard That*.

One way educators can support student efforts for holding polluters accountable is to familiarize students with map tools for understanding environmental data visualization. For example, ProPublica's Map of Cancer Alley can display what the estimated air toxicity levels from cancer-causing chemicals would look like once Formosa is running. This story map tool offers a vivid illustration of the health risks Lavigne warns about in her congressional testimony. Several of the tools listed in Table 4.6 will allow users to input an address or zip code to find how toxic the air is.

After exploring the mapping tools, ask students to document their findings in their own words. The information available through use of these tools can help students reconcile their notes about hazard indicators and firsthand accounts of community exposure. In the case of Cancer Alley, students can observe that residents are experiencing multiple, cumulative hazards at once (Younes et al., 2019). These tools strikingly convey what Rob Nixon has called "slow violence": the gradual but inextricable forms of violence against peoples and ecosystems (p. 2).

The following resources are indispensable aids in assisting students identify the decision-makers they wish to inform and sway.

Climate Action Now app: allows users to take quick action on important climate issues at the local, state, and national level right on their phone

GovTrack.us: publishes the status of federal legislation, information about your representative and senators in Congress including voting records, and original research on legislation

Websites for environmental justice groups: visiting websites or social media accounts for environmental justice groups can help people learn about the grassroots actions of communities and recommended actions to support their work

Modeling a "think aloud" as you project these resources on a screen will help students begin to imagine the possibilities that exist with these resources. For example, the Climate Action Now app lists environmental justice campaigns. Choosing "Clean Up Cancer Alley" presents users with a host of options for taking action. When visiting GovTrack.us, educators can show students

TABLE 4.6 Map Tools for Understanding Environmental Data Visualization

Name	What It Does	What Does It Help Us Understand?
ProPublica's Map of Cancer Alley, using the EPA's RSEI	Using the Environmental Protection Agency's Risk-Screening Environmental Indicators model, this ProPublica story map calculates the estimated chemical concentrations from toxic industrial plant emissions across the country, down to 810-by-810 blocks	Helps users find where toxic levels of cancer-causing chemicals are highest in the seven parishes found in Cancer Alley, Louisiana
AirToxScreen-Mapping Tool	Based on 2019 emissions, the Environmental Protection Agency's AirToxScreenMappingTool lets users "zoom" to areas of interest using the map's search tool and select census tracts to see summary level information about risk from air toxins.	Helps users learn total cancer risks for each U.S. census tract, cancer breakdown by pollutant and emissions type, emissions data for point-source sector emissions modeled in AirToxScreen, and air toxin monitoring sites with monitoring data
Climate and Economic Justice Screening Tool	In 2021, the Council on Environmental Quality developed the CEJST, which has an interactive map and uses datasets that are indicators of burden.	Helps users identify disadvantaged communities due to being overburdened and underserved
Treepedia	In collaboration with the World Economic Forum, the MIT Senseable City Lab developed a metric—the Green View Index—by which to evaluate and compare canopy cover.	Helps users compare cities and assess presence and absence of urban tree canopy cover

Source: ProPublica, Environmental Protection Agency, MIT Senseable City Lab

how to look up the voting record of a member of Congress on key environmental legislation. Exploring the website for Rise St. James, the organization founded by Sharon Lavigne, will let visitors know which actions they can take to best support residents, such as signing a petition asking JPMorgan Chase to divest and defund Formosa Plastics and instead make ethical investment decisions supporting healthy communities.

The And-But-Therefore (ABT) Storytelling Method

Learning how to sequence their research data into a story of *why* environmental injustices occurred and *how* they could be addressed meant practicing a new prewriting strategy: the "And-But-Therefore" (ABT) storytelling method. ABT gives writers a structure for plotting a story through its key developments: the situation, the complication, and the resolution. Olson explains that the ABT storytelling method is recognizable in many popular texts, ranging from the structure of Lincoln's "Gettysburg Address" to the plot formula of *South Park* episodes (p. 107).

As a mentor text, *Mayah's Lot* models how one person can compel others to collectively mobilize in an effort to protect the healthy well-being of one's community. After our initial reading of *Mayah's Lot*, I summarized the story using the ABT structure:

> A girl lives in Forestville, NY AND tends a garden on the lot off Lincoln and 121st, BUT she realizes a company plans to store toxic waste on the lot. THEREFORE she mobilizes community members to speak at the neighborhood council meeting to halt the company proposal.

This summary in ABT form distills the comic's narrative richness into discrete parts to suggest the structure of causal narratives:

> The "AND" joins two facts about the subject and presents a situation.
> "BUT" shifts the narrative direction due to an arising complication.
> "THEREFORE" heralds a consequence after a period of time.

When my students constructed their research presentations using these craft moves, the root causes of an identified environmental injustice became far more comprehensible.

Connecting to Other Environmental Justice Issues

Mayah's Lot taught my students the necessity of access to green spaces for mitigating harmful environmental impacts. As a class, we prepared for a community presentation on a local environmental injustice regarding urban tree canopy. When reading the panels that show Mayah's new friend, Troop, encouraging her to attend a community council meeting, my students wondered why Troop was wearing an oxygen tube. As they mulled possible reasons, another student was reminded of an area near Los Angeles International Airport described as "asthma town." Due to exposure to multiple pollution sources caused by air traffic and freeway commuter traffic, community residents living in South Los Angeles experience harmful health impacts. My students inferred that Troop's use of an oxygen tube and tank was possibly due to air toxin exposure in his overburdened community. This insight directed their attention to the comic's final panel, where the definition of environmental justice returns the reader to the Aspen seed imagery: "Just like the roots of the aspen, we're all tied to each other, holding our cities or towns together."

Indeed, it is the lack of green spaces to begin with that accounts for many unhealthy community conditions. Communities of color that have experienced what Sarah Milligan-Toffler calls "nature redlining"—a lack of parks, gardens, and green spaces due to racist housing practices from the 1930s—will feel the brunt of budget shortfalls for nature programs (Chiotakis, 2020). The studies my students consulted show that people who live next to city factories, plants, and transportation hubs suffer a higher degree of asthma and other respiratory diseases. Trees themselves play a major role in removing pollution from the air, as well as in filtering dust, which contributes to the onset of childhood asthma. Through looking at urban tree canopy measurements on *Treepedia*, we learned that cities near the Los Angeles airport show almost no green coloring, which means there

TABLE 4.7 Call to Action Structured by ABT

And	But	Therefore
Surrounded by the natural beauty of Mulholland Park, I wondered how more students could stand where I'm standing and have access to a much needed source of peace and quietude.	Without protected tree canopy, many LA communities will suffer increased incidence of respiratory diseases. Urban heat island and drought conditions will become more severe, even deadly.	Contact your representative to urge support for Assembly Bill 1530. We need urban forestry funding for a resilient, more sustainable future.

is very little canopy cover in these areas. As they began using the ABT method, students expressed the connection between income inequality and park-poor communities. One of the reasons that the LULUs and plants they researched are so harmful to Angelenos is their siting in communities that have fewer resources for tree plantings on public properties and a smaller tax base for tree maintenance. Table 4.7 exhibits the call to action my students invited audience members to take in the form of a postcard sent to our local representative. This illustrates how the ABT method can help students structure their calls to action and urge support for an Assembly Bill.

Mentor Text Moves for a Message of Advocacy

Selecting mentor texts to guide students' steps into advocacy does not have to fall on the educator's shoulders alone. Throughout the research investigation, encourage students to point out and save those memorable interviewee quotations, transcripts, speeches, and letters that offer models for crafting effective calls to action. The following excerpts can help educators present examples of powerful craft moves. Use the following questions to elicit craft noticings from students:

- Can you name what the speaker is doing?
- Why is the speaker doing this? What is it adding to the writing?

#1 ALLUSION

"Everybody remembers what happened in Flint. There's hundreds of Flints all across America. How many of you know, when you send your child to school, the fountain they're drinking out of is not fed by a lead pipe?" —Remarks by President Joe Biden on the American Jobs Plan, April 7, 2021

#2 RHETORICAL QUESTION

"Remember what a permit is. You know what a permit is? A permit is permission to pollute." —General Russel Honoré, Founder of GreenARMY

#3 PATHOS

"You smell Cancer/Death Alley long before you are in it. The odor changes depending on what the 150 chemical plants or refineries are releasing into air that day. It could smell like rotten eggs, burning chemicals or something even more pungent." —Tammy C. Barney, *Louisiana Illuminator*

#4 LOGOS/ANECDOTE

"My sister died at the age of 43 from an allergenic disease called sarcoidosis, a disease which affects 1 in 1,000 people in the United States, yet in Norco there are at least 5 known cases in fewer than 500 people of color." —Margie Richard's United Nations testimony about what it is like living next to Norco's Shell Oil Refinery

#5 SCARE TACTICS/ANECDOTE

"Every local community organization is approached by the refineries for funding. That's the way they co-opt people from speaking about how they are harmed by the activities of the refineries and the pollution (Moreno, 2022)." —Alicia Rivera, a Wilmington community organizer at Communities for a Better Environment

Conclusion

To support students as they become environmental justice activists, educators can target four essential skills: identifying factors that contribute to different risk burdens, communicating why environmental hazards are public health emergencies, identifying mechanisms for accountability, and advocating for the prevention of adverse health impacts. The ability to illustrate concepts and conflicts at the heart of these topics can demystify the ideas central to them. By using comic storyboards to explain local examples of environmental injustice, students are engaged in acts of rehearsal for conversations outside the classroom. They are also able to practice their storytelling before presenting their research to outside audiences in contexts with higher stakes.

Next Steps
The following questions can help students build on their research investigation and assess their findings from new angles.

> How can investment in infrastructure help prevent incidents of environmental injustice?
> What biases in news coverage can contribute to the downplaying of community action?
> How did the passing of the Bipartisan Infrastructure Law and Inflation Reduction Act offer new opportunities to support environmental justice grassroots organizations?

A Peek Into Their Practice

"You'll see more connections with trends and ideas if you don't get so caught up in the minutiae."
   Myah Lunceford teaches World History and Advanced Placement African American Studies at Grover Cleveland Charter High School in Reseda, CA. A year ago, she guided a group of teachers including myself through a tour of

historical documents about planned neighborhoods. Myah drew attention to the type of language used in some of these documents: "Why do you think this word was used?" I was eager to learn more about how she connected her urban planning background with environmental studies—how she helped students notice the way people map themselves could be both expansive and exclusionary.

Myah explains that maps are the key to understanding redlining. "Students look at what communities they live in looked like eighty years ago. Then they look at an overlay map and select a parcel to compare past mapping with contemporary mapping. The question that is then raised: How does this redlining compare with segregation that isn't as obvious today?" Students begin to understand how legislative policies can have a ripple effect.

When I asked Myah about how we can foster big lens thinking, she invited me to think about more general trends instead of getting caught up in so many particular details.

"We can talk about why an event, so long ago, still has implications today. We can base a lesson around determining who is most responsible for the atrocities committed in the Belgian Congo in the late 19th century, early 20th century. But we can also research mining practices for the cobalt used in our cell phones that occur in the Congo today and recognize that the extraction process and poor treatment of workers is connected to long ago events in the same place." Myah stresses recognition of the lingering effect of past abuses and underscores why they still matter in the present tense. "This isn't just numbers or dates. They are people. This is something that happened to people. When we keep that in mind, it's easier to bring in an environmental focus on inequities."

Surfacing Personal Connections

To help students understand the health risks inherent in physical proximity to environmental hazards, Myah invites students to study population maps with these questions in mind:

Who lives next to the freeway?
Who lives next to the airport?
How does this affect my family?

Guiding Question: What roles do historical maps and present-day maps play in helping us understand local environmental injustices?

ELA/Literacy Standards:

CCSS.ELA-Literacy.RI.9–10.7

Analyze various accounts of a subject told in different mediums (e.g., a person's life story in both print and multimedia), determining which details are emphasized in each account.
　　CCSS.ELA-Literacy.SL.9–10.1.c
　　Propel conversations by posing and responding to questions that relate the current discussion to broader themes or larger ideas; actively incorporate others into the discussion; and clarify, verify, or challenge ideas and conclusions.

Connections to Crosscutting Concepts:

Stability and Change: tracking urbanization through maps
Cause and Effect: tracking effects of transportation networks on population distribution

References

AirToxScreen. (2022, December 21). *AirToxScreen mapping tool*. Retrieved April 16, 2023, from https://epa.maps.arcgis.com/apps/MapSeries/index.html?appid=e5a7e59018c7424eaddcb64b31ba4a41

Barney, T. C. (2021, February 10). *You only have to smell 'cancer alley' to know how toxic it is*. Louisiana Illuminator. https://lailluminator.com/2021/02/10/calling-louisianas-petrochemical-corridor-cancer-alley-cant-be-a-slam-when-its-true/

Bates, J. (2019, August 27). Newark officials providing bottled water to 15,000 homes over lead contamination concerns. Here's what your need to know about the city's water crisis. *Time*. https://time.com/5653115/newark-water-crisis/

Biden, J. (2021). *Remarks by President Biden on the American jobs plan* [Speech transcript]. https://www.whitehouse.gov/briefing-room/speeches-remarks/2021/04/07/remarks-by-president-biden-on-the-american-jobs-plan-2/

Bratspies, R. (2019). Mayah's Lot: Teaching environmental justice with comic books. In C. A. Corcos (Ed.), *The media method: Teaching law with pop culture* (pp. 505–522). Carolina Academic Press.

Bratspies, R., & La Greca, C. (2015). *Art by Charlie La Greca. Mayah's Lot. Environmental justice chronicles: Book 1*. CUNY Center for Urban Environmental Reform. https://cuer.law.cuny.edu/?page_id=1272

Bullard, R. D. (2000). *Dumping in Dixie: Race, class, and environmental quality* (3rd ed.). Routledge.

Calenviroscreen 4.0. (2022, December 1). *OEHHA*. Retrieved April 16, 2023, from https://oehha.ca.gov/calenviroscreen/report/calenviroscreen-40

Chiotakis, S. (Host). (2020, June 10). Nature redlining' cuts off black and latinx families from nature [Audio podcast episode]. In *Greater LA*. KCRW. www.kcrw.com/news/shows/greater-la/nature-race-lausd-santa-monica/education-nature-race

Clearfield, C., & Tilcsik, A. (2018). *Meltdown: Why our systems fail and what we can do about it*. Penguin Press.

Climate and Economic Justice Screening Tool. (n.d.). *Climate and economic justice screening tool*. https://screeningtool.geoplatform.gov/en/#3/33.47/-97.5

Commission for Racial Justice. (1987). *Toxic wastes and race in the United States. A national report on the racial and socioeconomic characteristics of communities with hazardous waste sites*. United Church of Christ.

Coombs, D., & Bellingham, D. (2015). Using text sets to foster critical inquiry. *English Journal, 105*(2), 88–95.

Environmental Protection Agency. (n.d.) *US EPA environmental justice*. https://epa.gov/environmentaljustice

Fisher, D., & Frey, N. (2013). Making the most of graphic novels. *Engaging the Adolescent Learner*, 1–10. https.//doi.10.1598/e-ssentials.8007

Ganim, S. (2018, November 28). For 10 years, a chemical not EPA approved was in their drinking water. *CNN*. https://www.cnn.

com/2018/11/11/health/denmark-sc-water-chemical-not-epa-approved/index.html

Goldman Prize. (n.d.). *Goldman prize*. https://www.goldmanprize.org/recipient/sharon-lavigne/

Green, E. L. (2019, November 6). Flint's children suffer in class after years of drinking the lead-poisoned water. *New York Times*. https://www.nytimes.com/2019/11/06/us/politics/flint-michigan-schools.html#:~:text=the%20main%20story-,Flint's%20Children%20Suffer%20in%20Class%20After%20Years%20of%20Drinking%20the,children%20with%20high%20lead%20exposure

Greenaction. (n.d.). *Greenaction for health and environmental justice*. http://greenaction.org/what-is-environmental-justice/

Greenfield, N. (2022, November 10). *Advocates are sparking a revolution in Louisiana's cancer alley*. NRDC. https://www.nrdc.org/stories/advocates-are-sparking-revolution-louisianas-cancer-alley

Hughes, S., Dick, A., & Kopec, A. (2021). Municipal takeovers: Examining state discretion and local impacts in Michigan. *State and Local Government Review*, *53*(3), 223–247. https://doi.org/10.1177/0160323X211038862

Jackson, D. Z. (2014, October 22). Letter: Flint residents deserve better water. *The Flint Journal*.

Jackson, D. Z. (2017, July 11). *Environmental justice? Unjust coverage of the Flint water crisis*. Shorenstein Center on Media, Politics and Public Policy. https://shorensteincenter.org/environmental-justice-unjust-coverage-of-the-flint-water-crisis/

Jacobs, H. H., & Zmuda, A. (2023, February 1). *Storyboarding your curriculum* (Vol. 80, No. 5). ascd.org. https://www.ascd.org/el/articles/storyboarding-your-curriculum

Méndez, M. (2020). *Climate change from the streets: How conflict and collaboration strengthen the environmental justice movement*. Yale University Press.

Moreno, E. (2022, November 17). These moms are leading the fight against environmental racism. *The Nation*. https://www.thenation.com/article/activism/mothers-toxic-air/

Nixon, R. (2013). *Slow violence and the environmentalism of the poor*. Harvard University Press.

Olson, R. (2015). *Houston, we have a narrative: Why science needs story*. University of Chicago Press.

Rise St. James. (n.d.) *Rise St. James.* htttps://risestjames.org/

Storyboard That. (n.d.) *Storyboard that.* https://www.storyboardthat.com/

Treepedia. (n.d.). *Treepedia:: MIT senseable city lab.* http://senseable.mit.edu/treepedia

Vaughen, K. (2023, March 23). Michigan still dealing with fallout from Flint water crisis 9 years later; plus new water worries. *CBS News.* https://www.cbsnews.com/detroit/news/michigan-still-dealing-with-fallout-from-flint-water-crisis-9-years-later/

William, D. (2013). Assessment: The bridge between teaching and learning. *Voices from the Middle, 21*(2), 15–20.

Younes, L., Shaw, A., & Perlman, C. (2019, October 30). In a notoriously polluted area of the country, massive new chemical plants are still moving in. *ProPublica.* https://projects.propublica.org/louisiana-toxic-air/

5

Climate Stewards as Problem Solvers

Savvy consumers of media are not only good at fact-checking. They are also good at fact-finding. Climate stewards looking to increase their impact help others get to know their own representatives and how to influence them.

Educators are encouraged to design lessons that enable students to identify authentic audiences for their writing. Students need practice identifying the best audience for their reach: the best targets for their advocacy. Power mapping is a visual exercise that helps you to identify the levers and relationships you can take advantage of to gain access to and influence over your target.

Activity: To familiarize students with their representative, ask them to engage in some fact-finding.

When I introduce power mapping, I begin with a bodily metaphor—pressure points. Pressure points are parts of the body believed to be extra sensitive. When pressed, they can stimulate relief in the body. Asking students to identify the best possible targets with whom to press their message is a way to stimulate response where it is needed most. Power mapping gives students experience with being strategic with available resources, gauging the amount of time they themselves can devote to targeting

TABLE 5.1 Learning About Your Representative

Study district demographics. To whom is your representative accountable?
What bills (related to your concerns) has your representative authored or co-authored?
What was your representative's voting record on climate bills in the last session?
What are your representative's top issues and viewpoints?
What are your representative's committee assignments?
Who are influencers and key financial backers in their orbit?

efforts, and articulating insights about what motivates people to finally say "yes."

There are many ways we can look to influence our representative's vote:

- Sign online petitions
- Call or email the Capitol office
- Attend town hall meetings and speak out
- Find opportunities to thank your legislator and staff for their work

Planning Your Power Map

Sometimes it seems like there are too many potential targets, and it's difficult to winnow down the list. My students Kayla, Yijing, and Alondra identified many problems associated with food production, overconsumption, and food waste. Kayla realized many people had no idea about the landfill impact of food waste or how much misinformation could be traced back to diet culture. Yijing's research on grocery store layouts uncovered how aisle

TABLE 5.2 Identifying Possible Targets

Who is the best target for the change you are seeking?
What and who might influence them to say "yes" to your request?
How can you put pressure on the target?

and display layouts encouraged impulsive purchasing. Alondra noticed that most people don't know what farmers go through, citing extreme temperatures, unreliable crop yields, and seasonal disruption caused by climate change.

When initially researching potential targets, there are two important questions to keep in mind:

What really needs to change?
Where is that change made?

Taking them at face value, you can see both questions generate multiple answers. The more you learn about a problem, the number of root causes you are able to identify multiplies. At that point, it's easy to become overwhelmed. I find that bringing in the concept of "multisolving" can help students optimize their choices. Elizabeth Sawin and her colleagues began using the word multisolving to define this approach: "using one investment of time, money, or energy to address multiple problems" (2024, p. 15).

So I put a new question to my students: What is something we can do that addresses multiple problems at once?

As much as it seems easier to work on solutions alone or with those with whom we are already familiar, we need to work in broad partnership to meaningfully address the impact of our climate's most extreme changes. Whenever I research approaches to environmental problem-solving, one topic comes up again and again: silo reduction. Helping students work in unison, instead of isolated and apart, is one way we can prepare them to collaborate and mobilize with others. If your students are like mine, they understandably can feel ill-prepared by the magnitude of worsening environmental challenges. For this reason, I suggest linking discussion of climate solutions to the 17 Sustainable Development Goals (SDGs) set up by the United Nations General Assembly in 2015.

The goals offer a blueprint for peace and prosperity for all countries in global partnership. For example, by selecting the online icon for Goal Six: Clean Water and Sanitation, students can access information about events, actions, and publications geared toward making that goal a reality. Through considering the importance of water access in relation to the other goals, we can help students see how solutions intersect. This activity can help them better weigh the benefits of a proposed environmental solution. If weighing the feasibility and attractiveness of a proposed water supply solution includes thinking about the advantages it creates, then capturing stormwater could potentially reduce flood risks and protect fragile aquatic ecosystems, supporting Goal Fourteen: Life Below Water.

The SDGs offer a framework of collaboration that can sustain environmental problem-solving throughout the school year and beyond classroom walls. Revisiting some terms related to systems thinking, we can think of school as a system with many parts. The adjustments and overhauls needed to make it work

FIGURE 5.1 United Nations Sustainable Development Goals.
Image Credit: www.un.org/sustainabledevelopment (The content of this publication has not been approved by the United Nations and does not reflect the views of the United Nations or its officials or Member States).

optimally means addressing relationships between the parts and thinking holistically about interconnectedness. Combining skill-building in Critical Media Literacy and skill-building in climate stewardship is a way school leaders can leverage a school curriculum for maximum impact.

Large-scale climate solutions often have not considered the role that education can play. Improving Critical Media Literacy in schools leverages the web of networks that exist in a school district: from adoption of class texts to professional development opportunities to lessons centered on sifting through all the forms of idea and image provocation we encounter in the media ecosystem.

Helping students communicate as climate stewards has the potential to connect many literacy targets in schools: civic literacy, digital literacy, media literacy, and science literacy are just a handful of the literacies that can be folded into our focus on climate stewardship. We know students need frequent and varied learning opportunities to critically evaluate the validity of information they consume. The ease with which unverified information can spread through social media apps, podcast shows, and cable television channels attests to the necessity of helping students detect misinformation and disinformation. As professional development efforts focus increasingly on Sustainability and Critical Media Literacy, opportunities to target skills related to both areas multiply.

SDGs related to Critical Media Literacy:

- 4—Quality Education
- 12—Responsible Consumption and Production
- 13—Climate Action
- 16—Peace, Justice, and Strong Institutions

Activity: Reflecting on Your Own Media Literacy

To help students reflect individually on memorable ads that have lingered in their memory, invite them to reflect on influential visual media in their own life.

What are three ads that have influenced your awareness of a product or company? Explain each ad. Why do you think it was memorable?

For each ad you mentioned, find a related image online and bookmark the webpage.

What do you think it means to be a "media-literate" consumer? If someone is savvy about marketing and mass messaging, what might they be aware of?

Given the popularity of the ads you thought about, what techniques would you use if you were trying to design an influential marketing/public awareness campaign?

Ad Analysis Protocol

To increase critical media literacy, we can create a classroom environment where discussion is built around a familiar routine of ad analysis. Carolyn Fortuna (2016) offers a series of questions that can elicit interpretation based on scrutinizing specific elements:

> How are lighting, sound, music, voice-overs, special effects, editing, color symbolism, and/or casting used to foster audience interest?
> Is there any specific implied prior knowledge that would be important for a viewer to hold in order to understand the commercial? If so, name it.
> Describe the setting: time and place. Why did the composer choose these instead of other possible times and places?
> If there are people pictured in the advertisement, who is pictured? Who is left out and why do you think they have been left out?

This analysis protocol is useful because it can be used with historical and contemporary media, as well as popular culture texts. Students benefit from low-stakes opportunities to ask classmates for clarification about what they remember seeing in ads.

> ### A Peek Into Their Practice
>
> Vincenzo "Enzo" Loconte is a science teacher at Grover Cleveland Charter High School in Reseda, CA. The Los Angeles Unified School District formed the Climate Literacy Champion Program, a body composed of teachers who take on an additional role of supporting climate literacy at a school site. This program was how I discovered Enzo, who supports both students and educators in connecting what's happening on campuses with climate change solutions.
>
> "When it comes to climate change, my degree in psychology is probably most informing of behavioral changes that need to happen." Thinking about the way stories are told in our culture, he notices that "very seldom is the villain the system itself." The Thingamabob Game illustrates Enzo's point well.
>
> Developed by Bill Bigelow, the Thingamabob Game is a simulation game where groups of students represent rival manufacturers of goods. Bigelow and Swinehart's directions emphasize the stakes: "They will be rewarded based on how much profit they produce for their company, but the more thingamabobs they produce, the closer they bring the planet to climate catastrophe and environmental devastation" (2015, p. 148).
>
> While developing goods for profit, students don't know the precise trigger point of environmental destruction—the amount of carbon dioxide released into the atmosphere causing a catastrophic tipping point. "My students almost always go over that number. The game illustrates how a profit-driven economy is not aligned with environmental responsibility." Enzo sees the effect these discussions have on his students, who often find inspiration in stories of youth action and resistance movements that have had sizable changes. "Part of deciding who you want to be is figuring out what you feel most motivated by."

> Enzo explains one way educators can help students become aware of the effects of misdirection. "Public climate deniers use the same playbook as tobacco companies who attempted to minimize liability. If you keep a debate going about whether something's really happening, you're delaying action and not really getting to the point of solving it."

The Crying Indian Ad and the Corporate Tactic of Shifting Blame

The anti-litter organization Keep America Beautiful, in partnership with the Ad Council, launched the "Crying Indian" PSA Campaign in 1971. When a bag of trash is flung from a car and scatters near the Indian's feet, a tear rolls down his check while the narrator intones: "People start pollution. People can stop it."

A focus on Critical Media Literacy helps students parse out rhetorical elements so that they can understand how a message "lands" on the consumer as intended. We can help them assess media messages with a greater awareness of how their individual purchasing power is being targeted. Cultivating this awareness requires asking the question: How did the minds behind this ad want you to feel as you viewed it? Articulating the intended affective experience is key to understanding the choice of images present in the ad.

We can help students notice how our awareness of an ad's subtext stems from our ability to analyze visual rhetoric. One student describes the subtext of the Crying Indian Ad: "America is being tainted by trash, and it's our fault." We discussed how Indigenous people in this country tend to be regarded as having a more reverential relationship to the land and thus the person in the ad was being presented as an ideal example of how to be a climate steward. The ad was designed to deflect environmental responsibilities—the goal of the Keep America Beautiful marketing campaign was to shift culpability for plastic waste from the corporate manufacturers to the individual consumers.

Developing a habit of thinking about the connotation of a word or phrase and finding out through discussion if there is class consensus or divide regarding that diction choice is a great use of instructional time—students can dispel any misconceptions of words encountered in the texts we are scrutinizing. Students may gain clarity about the meaning of an abstract concept and may also gain experience in seeing how a word can possess multiple registers of meaning. To explore the connotations of the name of the Ad Council responsible for the Crying Indian Ad, I asked students to think about what the name "Keep America Beautiful" might hide about the Council's motives for creating the ad. Names having words with positive connotations can obscure harmful effects.

Zines

Some of my students pass the time watching YouTube videos of crafty things that they could do at home. They love learning about making zines: the tactile experience of creating a magazine in miniature allowed them to explore and document sources of passion or objects of curiosity. Because a zine is curated to reflect something meaningful to its maker, it seemed a fun way to foster environmental storytelling: my students and I would make mini-zines and share them with an audience of sixth graders at a nearby middle school.

We decided to make eight-page zines to choose an aspect of environmental stewardship we wanted to educate about, aiming for the clear, conversational brevity we associated with podcasts episodes found on *TILclimate Podcast*, and the podcast format of *How to Save a Planet*, where every episode ends with a specific call to action. Being able to describe an environmental problem in terms of its impact is vital because the impact can be difficult to discern. It is important to engage in storytelling that provides concrete ideas for taking action but is also accessible to a younger audience. This knowledge helps us adjust the scale of intervention our zine stories recommend.

Climate Stewards as Problem Solvers ◆ 143

FIGURE 5.2 Shows a page from a zine about gaming. My student Joseph's zine draws attention to the issue of electricity consumption and subverts the goals of PC company ads that wish to convince gamers that they need to buy more efficient PC parts. Instead, Joseph educates his audience about how to optimize game performance without spending more money. Joseph perceptively suggests an alternative—teach youth how to optimize game settings so that less power is used, circumventing the problem of phones and other devices overheating. Joseph's zine clearly illuminates a wider problem: a culture of consumption, based on a pervasive corporate tactic of convincing consumers that a current device will be imminently obsolete, leads to teeming landfills.

The example of the Little Free Library beautifully illustrates the ethic of reciprocity that lies at the heart of the gift economy. When writing about the serviceberry, Robin Wall Kimmerer explains the profound shift that happens when we see something as a gift instead of a commodity:

> The relationships nurtured by gift thinking diminish our sense of scarcity and want. In that climate of sufficiency,

FIGURE 5.3 Shows a page from a zine about street vendors, who work tirelessly in the heat, serving sweet and savory treats. My student Emma's zine draws attention to the working conditions of street vendors who sometimes encounter unwelcoming, even violent behavior. Emma's approach shows the intersecting concerns of environmental justice and social justice, creating a bifocal view of protecting the environment and people. The call to enact laws that will protect *los vendedores* challenges the audience to imagine how we can safeguard the health and dignity of people working in our communities.

> our hunger for more abates and we take only what we need, in respect for the generosity of the giver. Climate catastrophe and biodiversity loss are the consequences of unrestrained taking by humans. Might cultivation of gratitude be part of the solution?
>
> (Kimmerer, 2024, pp. 12–13)

In describing how the serviceberry moves through an ecosystem, endlessly transformed in a circular economy, Kimmerer points out how abundance, not scarcity, is achieved.

Climate Stewards as Problem Solvers ◆ 145

FIGURE 5.4 Teaches us that maps not only orient us; they reshape our mental terrain as we consider anew what is available to be enjoyed. A local Little Free Library, found on a street in our school's community, is a neighborhood delight. The concept of a little library, accessible and free to all, democratizes the ability to find a book and builds community through a book exchange. My student Tyler's zine zooms in on this Little Free Library, then proceeds to zoom out with each page, so that the wide-ranging network of Little Free Libraries across the globe is perceptible.

A Peek Into Their Practice

Rate: a ratio that compares two different quantities that have different units

Eric Wilson has worn a lot of "hats" in education. He's worked as a K–8 Technology Coordinator. He developed a STEM website for Twin Cities public television. He's worked as a STEM director at a large high school. His most recent teaching assignments were Algebra and Pre-Algebra. Currently, he works at TreePeople, the largest community-based organization conducting restoration and reforestation in Southern California, where he is working at schools converting hard-top playgrounds into green spaces. For much of my teaching career, he has been a generous thought partner, willing to collaborate on environmental projects and identify guest classroom speakers who can help my students think about sustainability in less abstract terms.

As someone who has occupied different vantage points from which to view the goings-on in classrooms, Eric sees clear opportunities to bring Math concepts into lesson design with a sustainability focus. "Students interact with rates all the time, but there's not a lot of tangible aspects to them."

"Think about everyday concepts. Why does this matter? What rates have we experienced?"

For students eager to learn how to drive, questions like "How many miles per gallon?" or "How many watts per mile?" invite genuine curiosity. "A lot of students have waited with a caregiver when the adult is filling up the tank. Rates invite practical considerations, like having to get a certain distance or mileage out of this."

Eric encourages students to take the reins. "Have them come up with examples. If they feel their voice is being heard, they are more prone to considering their own impact. I fear students are trained to tell the teacher the answer they think the teachers want to hear. As much as possible, make it relevant to them."

Guiding Question: How can we use rates to express reduced consumption?

Learning Standards:

CCSS.Math.Content.7.RP.A.1

Compute unit rates associated with ratios of fractions, including ratios of lengths, areas and other quantities measured in like or different units.

CCSS.ELA-Literacy.W.7.6

Use technology, including the Internet, to produce and publish writing and link to and cite sources as well as to interact and collaborate with others, including linking to and citing sources.

Connections to Crosscutting Concepts

Scale, Proportion, and Quantity: Expressing relationships between unlike units

Energy and Matter: Connecting rates of change to the transformation of energy and matter

Activity: Mass Awareness Campaign

How do young people become more civically engaged and increase their impact beyond the classroom? What are concrete actions young people can take? What are concrete actions young people can recommend to others to create a multiplying effect?

To tackle these questions, each group will create an awareness campaign devoted to reducing overconsumption in our communities. As part of the campaign, groups will create a media toolkit: online flyers with infographics, zines directed at a youth audience, and a ladder of engagement plan. This toolkit should offer clear recommendations about whom to contact about policies and legislation connected to reducing overconsumption.

To begin, read the online newsletter article created by Recycle Smart, an education initiative developed by the Massachusetts Department of Environmental Protection, called "Reframing the Waste Hierarchy" and learn more about the Eight Rs.

Box 5.1: List of Key Terms

The Eight Rs
Rethink
Repower
Refuse
Reduce
Reuse
Repair
Recycle
Rot

TABLE 5.3 Mass Awareness Campaign Rubric

	4	3	2	1
Connect SDGs to identify partners with common goals	Students make clear connections between partners based on SDGs	Students make connections between partners based on SDGs	Students make at least one connection between partners based on SDGs	No connections are made to identify partners
Create a Power Map to Identify Targets	Students create a map that clearly identifies well-chosen targets	Students create a map that identifies targets	Students create a map that identifies at least one target	No targets are identified
Communicates with Policymaker	Students communicate a clear call to action to policymaker/target	Students communicate a call to action to policymaker/target	Students communicate a call to action ineffectively	No communication is attempted
Implements a Ladder of Engagement Plan	Students create a plan with a clearly mapped-out network of engagement	Students create a plan with a mapped-out network of engagement	Students create a plan that insufficiently maps out a network of engagement	No engagement plan is made
Creates a Multiplying Effect with Impactful Messaging	Students successfully engage multiple audiences with their messaging actions	Students engage multiple audiences with their messaging actions	Students engage at least one audience with their messaging actions	No multiplying effect is created

Students must

- Clearly connect SDGS to identify partners with common goals
- Create a power map to identify targets
- Communicate with policymaker/target
- Implement a ladder of engagement plan
- Create a multiplying effect with impactful messaging

Conclusion

When we challenge ourselves to foster literacy practices connected to authentic purposes, we "break down the classroom walls" and build bridges to real-world contexts. When we give students opportunities to educate others, we are creating the multiplying effect we need to make as climate stewards who effect systemic change.

References

Bigelow, B., & Swinehart, T. (2015). *A people's curriculum for the earth: Teaching climate change and the environmental crisis.* Rethinking Schools.

Fortuna, C. (2016, February 5). *Let's analyze super bowl commercials.* Edutopia. https://www.edutopia.org/discussion/lets-analyze-super-bowl-commercials/

Kimmerer, R. W. (2024). *The serviceberry: Abundance and reciprocity in the natural world.* Simon & Schuster.

Sawin, E. R. (2024). *Multisolving: Creating systems change in a fractured world.* Island Press.

6
SEL and Climate Stewardship

SEL is the process through which all young people and adults: acquire and apply the knowledge, skills, and attitudes to develop healthy identities; manage emotions and achieve personal and collective goals; feel and show empathy for others; establish and maintain supportive relationships; and make responsible and caring decisions. If school leaders fear that climate stewardship will be perceived as "one more thing" to add to their staff's busy workload, linking it to a school's focus on social-emotional learning can help them make a case for its inclusion in curriculum design and development. The Collaborative for Academic, Social, and Emotional Learning (CASEL) delineates five broad and interrelated social-emotional learning competencies: Self-Awareness, Self-Management, Social Awareness, Relationship Skills, and Responsible Decision-Making. Lesson planning with these competencies in mind can anchor our work in supporting our students.

Exercises woven throughout the chapters in this book can help foster important characteristics related to responsible decision-making skills:

- Demonstrate curiosity and open-mindedness
- Can make reasoned judgements after analyzing information, data, and facts
- Identify possible solutions to problems

DOI: 10.4324/9781003517009-7

- Anticipate and evaluate the consequences of one's actions
- Reflect on one's role to promote personal, family, and community well-being

All these characteristics can help students reflect on one's role in promoting community well-being. The other SEL competencies draw more attention to emotional development on an individual basis, a much needed focus given recent survey results on young people's feelings about the climate crisis. A survey of 10,000 16-to-25-year-olds across 10 countries (Hickman et al., 2021) found "59% were very or extremely worried and 84% were at least moderately worried" about climate change. Part of my motivation for writing this book is to give educators the tools to make their classroom an identity-affirming space for would-be climate stewards—a space where learners and educators alike can safely explore and process emotions associated with climate change threats and uncertainties. This space can be built through combining social-emotional learning tools and writing exercises with the goal of tending to difficult-to-navigate emotions and unruly trains of thought—not pushing them away, but making space for them, and sitting with the discomfort.

Box 6.1: Key Term

Self-awareness: the ability to understand one's own emotions, thoughts, and values and how they influence behavior across contexts.

Activity: To begin planning how you will weave your SEL focus in your lesson design, consider administering a survey to your learners to help them identify how they are feeling. Student survey responses will help you gauge their perception of threats associated with the climate crisis and how you can devote resources and school time to addressing their emotions.

Which emotions do you experience when you think about the climate crisis?
Do you think we have a shared responsibility to take care of our planet?
Do you think we have collective rituals for mourning? Do we have space and time to mourn together?
Do you think humor can help people manage their fear and anxiety?
Do you think outdoor experiences can foster a sense of connection?

Focusing on what our students can do, instead of what they cannot do, is known as a strengths-based approach. This does not mean we minimize concerns—we still acknowledge that the challenging and difficult emotions associated with climate change need attention and support. But we utilize storytelling as a means to face the complicated feelings associated with this threat and to maximize a strengths-based approach at the school level. Instead of avoidance, we address challenges through teaching creative storytelling techniques that ask for both intellectual engagement and emotional engagement. Focusing on storytelling skill-building is the opposite of one-off SEL lessons that offer few opportunities for follow-through. This chapter weaves suggested writing exercises with concrete tools to support the emotional well-being of our students.

The Braided Essay

When it comes to recognizing and managing emotions related to climate anxiety and grief, it can be difficult to reconcile feelings that are uncomfortable and sometimes contradictory. Complex emotions cannot be easily shoehorned into tidy explanations, and more straightforward attempts to communicate these emotions may fail to offer a pressure valve. For this reason, I recommend a writing exercise with an unconventional structure that invites difficult-to-wrangle emotions.

The image of the braid is powerfully suggestive of attempts to reconcile incongruous threads. In this way, the braided essay can be a helpful teacher: an exercise in creative nonfiction that encourages nonlinear storytelling. Three narratives are brought together by connecting words or images that put the threads into conversation with each other. This can be a refreshing change of pace in the secondary classroom, where so much essay writing instruction is built around the five-paragraph essay rarely seen outside schools.

The memory images so faded they appear to be edged in sepia, the echo of a character's voice from something dear and dog-eared, the conversation fragments still playing in our heads on an unpredictable loop: bringing the flotsam and jetsam floating around in our minds into contact can produce new stories of self. In their book *Beyond Literary Analysis*, Allison Marchetti and Rebekah O'Dell (2018) describe how we can help students explore ideas by inviting them to sort evidence by categories. I love this suggestion because it asks students to identify connections while looking for footprints. The braided essay offers a similar opportunity by asking students to examine memories in a deliberately roundabout way. The braiding paves the way for exploratory writing that can help them see every thought and image as a new, possibly fruitful connection. What may appear off-topic or loosely connected in a different type of essay is seen here as worthy of further contemplation.

Activity: Before writing braided essays, study these three mentor texts. We look at Brian Doyle's essay (2019) to study explicit craft writing moves in prose, while we look at Heather Swan's poem (2020) and the myth of Icarus to study the use of structure and allusion:

Brian Doyle's "Joyas Voladoras"
Heather Swan's "Pesticide VII: Victor"
Icarus Myth

The stated parameters for writing the braided essay are as follows:

- At least one of the braid narratives should be personal and involve details from memory

- At least one of the braid narratives should include factual information gleaned from research
- Use at least three mentor text "moves" we've studied together (from "Joyas Voladoras," "Pesticide VII: Victor," and/or the Icarus Myth) to help you build your narratives
- Use connecting images, words, ideas, or even events that can get these narratives to speak to each other as you braid them

Mentor Text Move: Embed an Allusion

In order to escape imprisonment, master craftsman Daedalus created two sets of wings for himself and his son, each made of feathers glued together with wax. Though his father warned him not to fly too close to the sun, Icarus soared higher and higher, until the wax started to melt. Tragically, Icarus tumbled into the sea and drowned.

> The purpose of a vacation is to forget all the things that cause us grief in the real world. Going to bed late one night and waking up early only to find a mountain of work that needs to be finished within the next few hours leaves me tired, stressed, nauseous. The will to go to school slips further with each passing day. Senioritis is real. While my teachers go on teaching until the end, I painfully hold on by the end of their ropes. A long, overdue vacation that brings the images of crystal clear waters into reality is the only remedy to this senior-specific illness. Swimming through cool seawater under the beating sun—light, relaxed, free. Giving the same illusion of gravity-lessness as soaring through the sky on artificial wings.

My student Kayla cleverly weaves in an allusion to Icarus and his fateful flight as she ponders the transition from school year to summer vacation. By planting images and lines that will be echoed later in her braided essay, she contemplates her own coming-of-age story and how the prospect of a vacation trip with restorative "crystal clear waters" will be the cure for senioritis. These echoes work like a musical riff, creating an expectation that you know where the writer is going. By incorporating the image of one "soaring through the sky on artificial wings," she invokes the Greek myth, which gestures to the price we pay if we ignore words of warning. The common hope of youth is to revel in euphoric experiences; however, the story of Icarus contains a very clear message about the pain we incur when transgressing the limits placed on us. The common experience of human suffering, so vividly expressed in mythology, is undeniable, but my student's variation on the theme will place the suffering of nonhuman creatures center stage in the next "braid" of her essay. Playing with different meanings of "the end" allows her to acknowledge the weight of senioritis and the anticipation of freedom that a vacation will finally grant her fatigued brain and body.

Mentor Text Move: Use of Gerund Verbs (-ing Verbs) in Parallel Structure

> For hummingbirds came into the world only in the Americas, nowhere else in the universe, more than three hundred species of them whirring and zooming and nectaring in hummer time zones nine times removed from ours.
>
> ("Joyas Voladoras")

Doyle's essay is one that can be revisited endlessly for its delightful consideration of the hummingbird. As the reader is led through a maze of facts about this "flying jewel," they begin to realize that Doyle is talking about more than one thing at once: the hummingbird's heart; the size of a human heart compared to the interior chambers of the blue whale's heart, the question of how one will spend each of their heartbeats in their lifetime. My students and I discuss how skillfully Doyle's essay stages an animal encounter as an opportunity for self-confrontation. The roundabout path to the question of how one will spend their heartbeats was mapped out by the careful layering of facts, extended analogy, and use of repetition.

The Great Barrier Reef: a place of limitless beauty and life. The Great Barrier Reef: a place of great exploitation. Consider the coral: fringing, barrier, atoll—brimming with countless amounts of colorful life, reduced to a mass of stark white sorrow. The changing temperatures of ocean waters diminishes it down to its exoskeleton, an embodiment of its former self; a distant memory. They exist within a space, but can no longer sustain life. The purpose of their existence: no longer to house the species of the sea, but rather to serve as an attraction for a multitude of tourists. The beautiful water of the Coral Sea that houses the biodiversity of thousands, all floating and splashing and dying without a word, nor a sound. Icarus's untapered ambition sent him plummeting into the sea while our boundless ambition has us diving head first into humanly uninhabited waters, where the homes of hundreds of thousands aquatic creatures are subjected to the eyes of those who destroyed it for luxury. When the expectations of school buries me under mountains of stress and I can't go on, I think of the creatures swirling and twirling and weaving through the coral reefs.

Box 6.2: Key Term

Keystone species: an organism that helps hold an ecosystem together. Without its keystone species, ecosystems would look very different.

This braid of Kayla's essay is freighted with an anguished description of the world's largest coral reef system, where more than 40% of corals were killed off due to rising ocean temperatures in 2024. Kayla's description of the decimated ecosystem as being "reduced to a mass of stark white sorrow" draws attention to the duality of the Great Barrier Reef: it can be a lure for tourists looking for an escape while being transformed into a mass graveyard for coral colonies that could not withstand record-breaking ocean heating in 2023. The use of gerund verbs—"floating" and "splashing" and "dying"—emphasize both the vitality and the vanishing of a keystone species that ensures the balance of a delicate, precious ecosystem.

Mentor Text Move: Use a Line From One of the Mentor Texts as a Sentence Starter

The braided essay is often woven by threads representing the past, present, and future. It occurred to us as we read and studied Heather Swan's poem "Pesticide VII: Victor" that the present of the poem registered the process of disappearing—how bee populations are declining as a result of toxic pesticide chemicals. "Victor" is the name of a common pesticide spray; by giving the name to the poem, Swan invites the reader to think about the cognitive dissonance involved with linking victory with a chemical agent that contributes to the decline of our precious pollinators.

> "Pesticide VII: Victor" by Heather Swan
> The handful of dead bees
> she finds after the spraying
> are not the worst part
> for the beekeeper.
> It's the bees still struggling
> that gets to her. Limping
> in a circle like someone
> who's been spinning
> on a tire swing for too long,
> who then stands—dizzy,
> nauseous, stunned.
> Their wings shudder,
> but they cannot fly.
> These insects whose bodies
> know the rhythm
> of the blossoms,
> the changing angles
> of the sun, whose alchemy
> gives us liquid gold,
> whose love affairs

with pistils and stamens
give us apricots,
almonds, melons.
To witness is to be
dredged, she thinks.
What war do we think
we're winning?

Thousands upon thousands upon thousands of creatures left for dead. Thousands upon thousands upon thousands who had no say in their inevitable demise. And then I forget. Who has time to change the world in high school? I'm on vacation. To witness is to feign ignorance, she thinks. When did we start enjoying the beauty of life at the expense of the dying?

The last braid of Kayla's essay adopts the structure of the ending of Swan's poem. There is no tidy resolution to a recognizable dilemma faced by my student and her classmates, on the cusp of graduation, who may be looking forward to vacation plans but face ambivalent feelings about engaging in tourist activities in places ravaged by climate change. Writing braided essays reliably offers students a relatable way to explore avenues of parallel emotions. For many teachers, personal narrative writing is reserved for the beginning of the year, when we are trying to connect new faces with new names. As seen with Kayla's writing, giving students the opportunity to express stories of self-identity in an exploratory, experimental manner can make space for dissonant notes at any time.

Box 6.3: Key Term

Self-management: the ability to manage one's emotions, thoughts, and behaviors effectively in different situations, and to achieve goals and aspirations.

SEL and Climate Stewardship ◆ 159

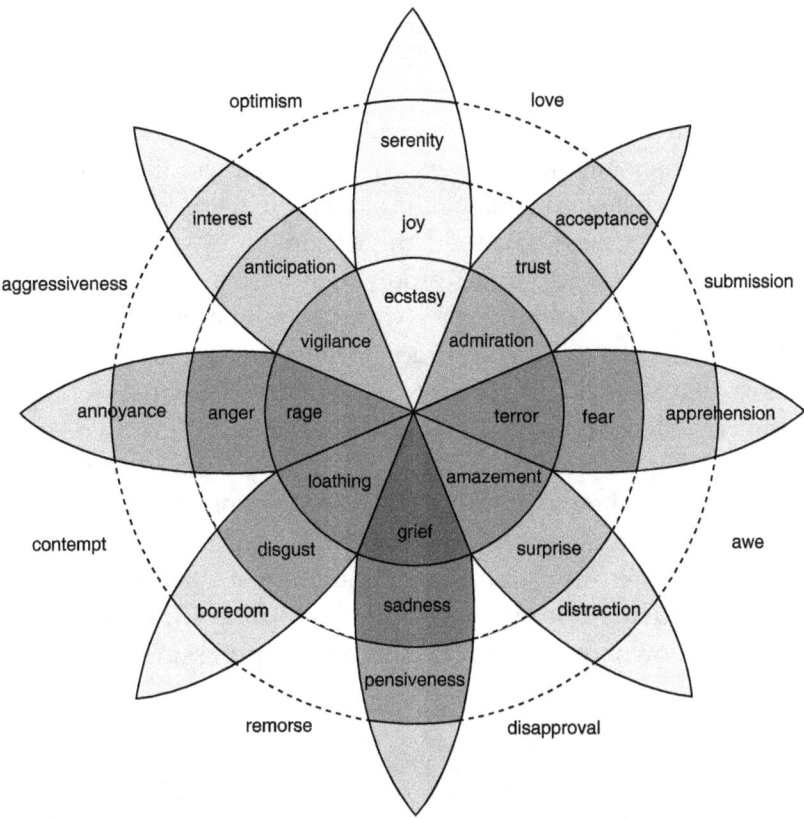

FIGURE 6.1 Plutchik's Wheel of Emotions. Public Domain.

A writing exercise like the braided essay can surface emotions difficult to explore. Psychologist Robert Plutchik (1980) developed a wheel of emotions that can serve as a helpful visual aid. He broke down eight primary emotions into opposite pairs: joy and sadness; trust and disgust; fear and anger; surprise and anticipation. While emotions near the center are more intense, emotions furthest from the center are mild.

Introducing students to Plutchik's Wheel can help them foster an important SEL competency: self-management. One way educators can support self-management is to help students become aware of their emotions in order to guide decision-making, as well as identify and use stress management strategies.

By embedding a writing routine dedicated to helping students become aware of and name difficult feelings, educators can pave the way for thinking about how to manage one's feelings. In her book *Social Emotional Learning and the Brain*, Marilee Sprenger (2020) explains that labeling an emotion can lower the activation of the amygdala while raising the activation of the prefrontal cortex, which aids with emotional regulation.

As soon as I read Kayla's ambivalent words about a vacation escape, I began searching for a way for my students to express what it feels like to hold contradictory feelings. Enter: Vent diagrams. Creators Elana Elsen-Markowitz and Rachel Schragis define a Vent diagram as "a diagram of the overlap of two statements that appear to be true and appear to be contradictory." On their website, they explain how Vent diagrams "help us recognize and reckon with contradictions" and "capture an ultimately unresolvable tension" (n.d., para. 1). Making a Vent diagram can liberate students from the pressure of adopting either/or thinking, where owning one feeling can mean suppressing another.

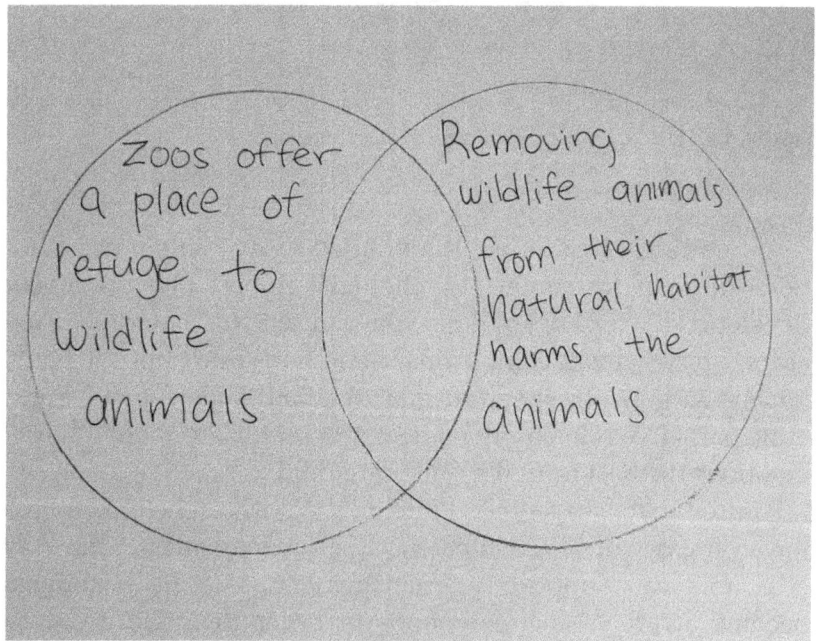

FIGURE 6.2 My Student Kingsten's VENT Diagram. Photograph by the author.

Activity: To underscore the benefit of labeling emotions, invite students to study Plutchik's Wheel and make their own Vent diagrams.

Flash Fiction

Though very short stories have been around for a long time, a new name for them—flash—was coined by James Thomas in his 1992 work, *Flash Fiction: Very Short Stories*. Thomas defined flash fiction as a story between 250–750 words in length. These "short short stories" depend on economy and efficiency, eschewing detailed backstory and description. When I began to teach flash, I noticed how much easier it became to set up open mic readings in class. Every student had a few flash pieces to share. Because the pieces are short, it was easier to be brave and volunteer to share one's writing. Interestingly, my students were developing their listening stamina, too. We easily could read a handful of flash pieces aloud in one class period. Since most flash stories are published in literary journals, we gained familiarity with online publications and a wide host of contemporary writers. Despite presenting a story world in miniature, flash was expanding our horizons.

The possibilities for teaching with flash fiction multiplied when a 2018 NPR article ("Butterfly Preserve on the Border Threatened By Trump's Wall") caught my eye. After we read about the unwelcome border wall construction on the 100-acre butterfly center—home to endangered plant life and trails that are the natural habitat of more than 200 species of butterflies—I invited my students to write a flash ghost story set at this nature preserve. To do so, I encouraged them to think about the experience of nonhuman animals in this disturbed natural setting and to amplify details related to haunting and loss uncovered in the news article.

I observed my students figuring out how to depict versions of being haunted as a result of some violence that has occurred, all in the compressed space of a flash story. One writer immersed his readers by giving us a "butterfly's-eye view" of what the butterfly is experiencing in the nature preserve, be it serenely floating by or seeking nectar from a sweet mock orange flower (a true favorite of our winged pollinators). The butterfly's home is then disrupted by a tractor upending ancient trees with teeth lifting into the air. He deftly moved from a description of a single tractor appearing monstrous to a collective image of a row of tractors achieving brutal demolition of the butterfly haven. Beset with confusion and uncertainty, the butterfly is forced to flee as it mourns the loss of its home and the home of its ancestors.

By recasting the details of a news article in the form of a flash ghost story, my student was able to process its implications in the form of storytelling, rendering this disruption with notes of empathy and grief. As we discussed the article about the threatened butterfly center, we thought about what it meant to consume the news with the counterstory stances in mind. My students knew discussing the border wall was a politically charged topic, but they were unaware that walls erected in the name of national security could have significant consequences for biodiversity. Looking at the wall through an environmental lens means considering how walls and fences built on the borderland conservation hotspots crisscrossing the U.S.-Mexico border can threaten a species' ability to maintain a migration route or to navigate an increasingly fragmented and degraded habitat. When we consider the consequences of human disruptions to landscapes, we think about both the visible and invisible signs of species and habitat loss. Writing ghost stories can help us acknowledge the way landscapes are haunted by the presence of nonhuman animals wrested from their homes due to human interference and building encroachment.

To help your learners get familiar with writing environmental ghost story flash, I encourage you to ponder some staples of the ghost story genre and generate a list of ways we define the words "ghost," "house," and "haunting." In Table 6.1, find some of the definitions that have been most fruitful for flash brainstorming in my classroom.

Activity: Defining Staples of the Ghost Story
Which of the definitions extends your students' thinking? Ask students to produce their own definitions of the same words and

TABLE 6.1 Features of a Haunted House

Ghost	An echo
	Something that disrupts the world of sense
	Something that lingers
	Disembodied spirit with a purpose
House	Family
	Shelter
	Time capsule
	Physical space charged with living energy
Haunting	A memory you can't ignore
	Something that refuses to be forgotten
	A wound that won't stop bleeding
	A barrier to peace

reflect on how the less literal, more figurative definitions push the boundaries of our thinking about what it means for a house to be haunted.

The elasticity of the ghost story—the way a ghost story can stretch our sense of boundary between the living and the non-living, past and present—makes it a valuable resource for contemplating the permeability of boundaries and the factors accounting for such spectral blurriness. When students write ghost stories, they are playing with clues and traces: they are figuring out how to incorporate clues gesturing to what has happened, as well as how to embed traces of a vital presence that once inhabited a now disturbed place.

Activity: Identifying Mentor Texts for Flash Stories

To help students contemplate how stories of loss and disappearance contain the vestiges of what is no longer here, invite students to read a constellation of flash stories trafficking in ghostly echoes. Far more subtle than a camp horror jump scare, the following stories model storytelling moves that can aid students in crafting their own environmental ghost stories:

"On Rannoch Moor" by Audrey Niven (2021)
   "It's Shaped Like a Grin, They Say" by K.C. Mead-Brewer (2018)
   "Eulogy" by Dina L. Relles (2020)

Together, these flash stories cleverly model how to build in an implied ending, invoke the oft-repeated details of an urban legend, and employ an object that comes to symbolize everything a life can represent.

Storytelling can help students frame things in a way that helps them make connections between what they know and what they are discovering—crafting ghost stories containing recognizable genre features that aid in making new story elements accessible. Eager to provide students with more opportunities to process the implications of the informational texts we read, I turned to flash fiction writing. Writing flash helped us contemplate the enormity of this environmental plight and reinforced the importance of sharing stories to engage emotion and empathy when mulling scientific information.

After the successful foray into flash writing based on the NPR butterfly preserve article, I continued to seek out other informational texts that would be a good fit for flash writing. The phrase "a message in a bottle" echoed in my head as I thought about how an environmental ghost story could serve as a type of call to action. The phrase conjures an image of a weather-beaten bottle, bearing a message from an earnest sender, floating among marine debris.

Activity: To begin this activity, read the online *National Geographic* encyclopedic entry about the Great Pacific Garbage Patch with students.

After we read the encyclopedic entry, the ghastly images of the marine debris forming the garbage patch took hold. Thinking about the "cloudy soup" of microplastics and abandoned fishing gear alongside the sweet, sentimentalized image of "a message in a bottle" reminded me of a writing exercise I learned from flash writer extraordinaire Kathy Fish (2020) about juxtaposition. She describes how poets and mosaicists deftly jam two or more different images together, which can allow something surprising to emerge. Following her example, I invited my students to choose one item from List A (all words taken from the *National Geographic* entry we read) and one item from List B, and combine them in a flash story no longer than 100–200 words:

TABLE 6.2 Combining Flash Fiction "Mosaic Pieces"

List A	List B
Gyre	Postcard
Garbage Patch	Ghost
Plastic Water Bottle	Photograph
Seal	Boat
Sea Turtle	Vacation

This juxtaposition exercise elicited flash gems—stories that glimmered as we held them up for inspection. My students always lean forward and look alert on mentor text study days, even more so when they know their classmates' writing will help us build our own repertoire of writing craft moves.

In her story, my student Alexa combined "sea turtle" and "ghost" to craft a domestic scene about a loggerhead sea turtle and his wife wishing to celebrate his promotion with his favorite meal of shrimp fried jellies. She draws on a specific detail included in the *National Geographic* entry: "Loggerhead sea turtles often mistake plastic bags for jellies, their favorite food."

What begins as a charming tale about anthropomorphized sea turtles morphs into a poignant statement about the sea life who become innocent victims of reckless waste disposal.

> Donny, the loggerhead sea turtle, was taking a trip to the grocery store. He was in a particularly cheerful mood, as he had been promoted to head turtle where he worked. Donny suddenly remembered the grocery list his wife had given him. She planned to make Donny's favorite meal; shrimp fried jellies. He went to retrieve the list and as he came back outside, he noticed something rather peculiar; the water was cloudy. Not wanting to worry his wife, he resisted the urge to call for her and kept this observation to himself.

****

As they sat at their makeshift dining table, he noticed that the jellies looked different than usual. He was speaking when a most alarming voice startled him. "Do not eat the jellies! You and your wife will fall ill if you do!" Donny did not believe in ghosts. He brushed aside the warning and dug into his food.

Writing Noticings

- "Smart surprise"
- Use of supernatural aid
- Unheeded warnings

Students loved how Alexa created an unexpected turn of events, shifting the reader from a place of comfort to discomfort. They appreciated that the ghost was not stereotypically spooky and wondered if Alexa was influenced by the description of "ghost fishing" (what was described as seals and other mammals drowning in abandoned plastic fishing nets in the *National Geographic* entry). Her flash helped us consider how frequently we ignore warning signs, a question we need to ask more often given that so many actors continue to evade responsibility for marine debris cleanup.

One student commented that more straightforward environmental appeals strike him as heavy-handed and preachy. As soon as something sounds like a lecture, he tunes it out. But his classmates' flash pieces hooked him—he understood the pathos of the situation and had greater appreciation for environmental organizations that were attempting to address the garbage patch while political actors were passing the buck. Flash writing can help students process the contents of an informational text as well as help them parse out and zoom in on details that warrant further contemplation. This type of writing exercise can also help students develop strong social-awareness skills, such as considering the perspectives of others.

In historical examples of famous "message in a bottle" incidents, the messages operate as love letters, distress signals, or memorial tributes. When reading this flash inspired by the Great

Pacific Garbage Patch, we can see all three motives combine: To treasure that which we love about the global ocean and marine life. To sound the alarm about the immeasurable damage we have caused. To appreciate what we have before it is gone.

The potential for schools to participate in this transformative work hinges on what they do to help channel private grief into the public work of mourning. Britt Wray, a scientist and expert on the psychological impacts of the climate crisis, identifies how grief and mourning differ: "Grief is a private undertaking that can exist solely in one's inner world, while mourning is the outward expression of grief. What we choose to mourn shows what we choose to value" (2023, p. 207). When we render ecological losses visible in shared stories, we acknowledge the stakes of our *collective* story. When we feel bereft of public mourning rituals, writing about environmental ghosts and experiences related to loss can help us address a key SEL competency: social awareness.

> **Box 6.4: Key Term**
>
> Social awareness is the ability to understand the perspectives of and empathize with others, including those from diverse backgrounds, cultures, and contexts.

A challenge to talking about ecological grief is it does not accommodate common narratives of what grief should look like. People will often avoid talking about grief because many of us are uncomfortable talking about things we cannot fix or cannot solve. This topic avoidance can lead to feeling like our emotions related to loss are being downplayed and diminished. When it comes to addressing ecological loss, we also contend with the effects of what wildlife biologist Whitney Kroschel describes as shifting baseline syndrome, which occurs "when each new generation perceives the environmental conditions in which they grew up as 'normal'" (2019, para. 3). An alternative to normalizing wide-scale species and habitat loss is to focus on kinship with nonhuman animals and to define "our baseline for a desirable state of our natural world." Cultivating writing skills in our classroom with this storytelling stance in mind—storytelling for reciprocity and kinship—can transform school sites into places of memorial and mobilization.

The Hermit Crab Narrative

Lately, I have been inspired by an unconventional approach to climate messaging: employing humor and playfulness. Sarah Yeo, a science communication researcher at the University of Utah, considers how scientific information can threaten people's existing beliefs and worldviews and the role humor plays in benignly conveying the same information, enabling greater receptivity (Thompson, 2022). Comedic approaches can help us process and retain information, especially since humor can lower the audience's guard. When we create conditions for playful experimentation, we can lower the stakes for communicating about a serious topic. In fact, we may lower an entire drawbridge, allowing students to enter into an imaginative space previously regarded as a formidable realm, where they can explore materials for arming themselves and disarming themselves as they wish.

This type of safe space brings to mind other creatures who armor themselves. Consider the hermit crab. Born without protection for its soft, exposed abdomen, the hermit crab spends its life inhabiting empty shells abandoned by snails and other mollusks. In honor of these perpetually shell-seeking creatures, Brenda Miller and Suzanne Paola dubbed a particular form of lyric essay the "hermit crab" essay—a type of essay that appropriates an existing form. By exploring material that is "soft, exposed, and tender," Miller and Paola explain, a writer may "look elsewhere to find the form that will best contain it" (2012, p. 128).

Familiarizing students with the text structure of the borrowed forms builds confidence. Instead of feeling like a writer meandering helplessly, the model text acts as an invitation to be playful, but within given parameters.

To help students weigh possibilities existing with borrowed forms, I made a list of familiar forms from which I sought mentor text examples.

Box 6.5: Form Choices

Possible Forms to Borrow
Recipe
How-to Article
"I regret to inform you" Rejection Letter
Crossword Puzzle
"Scale of 0 to 10" Medical Pain Scale
Math Problem
Encyclopedia Entry
Real Estate Ad
Dating Profile

Hermit Crab Essay Mentor Texts

How-to Article: Lorrie Moore's "How to Become a Writer"

In her debut story collection *Self-Help*, Moore (1985) included a series of "How-To" pieces, employing the step-by-step style used in how-to articles to narrate a writer's journey in a delightfully idiosyncratic unfolding.

The seminar doesn't like this one either. You suspect they are beginning to feel sorry for you. They say: "you have to think about what is happening. Where is the story here?"

Though the form of the essay sets the reader up to expect a clear, upward trajectory, we soon come to realize that each step offers insight into the frequent experience of emotional deflation that comes with receiving feedback on one's writing. Instead of offering a portrait of sure-footed success, Moore's essay suggests the choice to become a writer serves up a different experience with inevitability: the realization that *not* writing would be more painful than fielding the frequent underwhelmed responses to her career choice.

Rejection Letter: Brenda Miller's "We Regret to Inform You"

In an essay published in *The Sun Magazine*, Miller (2013) offers a series of rejection letters, adopting the voice of detachment that is so representative of this type of correspondence:

> Dear Young Artist:
> Thank you for your attempt to draw a tree. We appreciate your efforts, especially the way you sat patiently on the sidewalk, gazing at that tree for an hour before setting pen to paper, the many quick strokes of charcoal executed with enthusiasm.

The letters are dated and sequenced for maximum effect, beginning with humorous addresses to her younger selves that elicit chuckles, which make way for letters that divulge more serious, heartrending topics. Somehow, though, the detached tone of the letters sustains a humorous thread, perhaps arising from a voice steadfastly devoid of emotion. The form of the rejection letter becomes an experimental playground where a writer can engage in deadpan confession.

Real Estate Ad

My students love seeing how I approach the task I assign to them, so I created a hermit crab essay with a decidedly meta bent: I made a mock real estate ad directed towards a hermit crab seeking a new home. I had come across a disturbing Smithsonian article describing how hermit crabs are using trash as shells, painting a sobering picture of how human trash disposal is transforming the natural world.

> "A Spe-shell Place to Hang your Hat"
> This bottle cap shell offers a turnkey tiny house that blends modern convenience with functional living. Surround yourself in nature and soak in the sweeping ocean view with the remodeled open layout. Schedule a private tour today!

The compact nature of the real estate ad—you want to capture the reader's attention quickly—lends itself to bubbly hyperbole that in this case both reveals and conceals the devastation posed by the tiny toxic particles introduced to these organisms' bodies through degrading plastic. Tragically, many hermit crabs who adopt these artificial homes as protective havens find themselves unable to escape them and starve to death. Much like Miller's borrowed form, the mock real estate ad demonstrates how a serious topic can be explored in the guise of playfulness.

Kinship and Place-based Learning
Over time, educators can witness our students experiencing deeper kinship with the more than human world. During the COVID-19 pandemic, my student Maddie began her involvement with the Aquarium of the Pacific as a part-time volunteer over Zoom. Since then, her volunteering has evolved, and I sought to learn from her insights, as she occupies a unique vantage point where she can observe how aquarium guests interact with marine life and exhibits. Maddie has learned that people need gentle encouragement to move beyond initial discomfort. "Some people are afraid to touch Moon jellies, thinking 'Oh no, it's a jellyfish, it's going to sting me!' But they have really small stingers; they just don't have enough energy to hurt us. We ease them into the experience. 'Just start by touching the water.'"

Volunteering at the aquarium has given her experience with speaking into a microphone in front of 50 people, attempting to balance humor with strictness ("You can't jump in the pool; please don't do that.") Stoking respect and enthusiasm for marine life some guests have never heard of before, she is grateful for the aquarium's interactive experiences that encourage people to consider their own impact on delicate marine ecosystems and lay the groundwork for connection. "When people have hand-on experiences," Maddie explains, "they think about direct impacts and form an emotional attachment. This helps them think about impacts on a larger scale."

Maddie's emphasis on the value of tactile experiences is confirmed by research on the benefits of immersive, place-based learning. Studies (Bratman et al., 2019) have demonstrated a link between experiences in nature and increased psychological well-being, suggesting that time spent outdoors enhances memory, attention, creativity, and imagination. Drawing attention to the resources of sensory observation—focusing on what we know through hearing, touching, seeing, smelling, and tasting—has helped me connect SEL with a joyful exploration of our planet home.

To encourage students to spend time outdoors, design lessons that incorporate the use of plant identification apps for tracking their observations of the natural world. I remember the delight I felt when I first saw other users confirm my plant identifications and add my shared images to scientific data repositories on the iNaturalist app. By comparing your uploaded image with images of the same organism uploaded by other users, you are given the opportunity to see it from new angles—how a plant displays new colors during its flowering season or how a lizard changes color prior to skin shedding. The app mapping will display where observations have been recorded across the globe, providing insight into where a species is likely to flourish and where it is disappearing.

> Activity: Please download one of the following apps, and use the app to begin identifying plants.
> iNaturalist
> PlantNet
> PictureThis

Your goal is to identify a native plant that you can observe over a two-week period. Are leaves emerging? Are flowers blooming? Is fruit ripening? Carefully study your native plant and make notes on the first date and last date of observation.

Box 6.6: Key Term

Phenology: the study of how seasonal and climate variations affect the life cycle of plants, animals, and microbes

---

SEL and Climate Stewardship ◆ 173

Observing over time allows students to appreciate subtleties unnoticed by many. For me, repeated visits to the same observation spot have made me alert to even the smallest changes to a plant's appearance, stretching my vocabulary to pinpoint the differences I observe. In her book *Poemcrazy*, Susan Goldsmith Wooldridge (2009) describes how her own strong gathering instincts have made her a collector of words. While others may collect shells or stamps, she lists words in her journal, never knowing exactly when one might trigger a new poem. To this end, she tosses them into a figurative "wordpool."

Activity: A "wordpool" is a wonderful visual representation for how words combine and swirl together. Create a "wordpool"

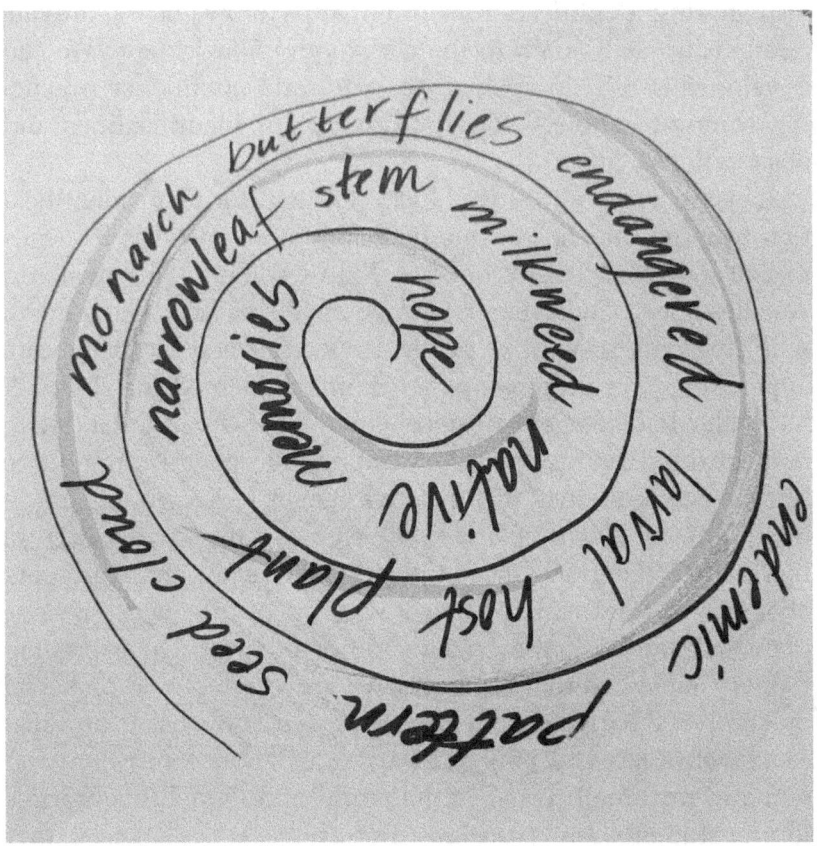

FIGURE 6.3 Image of "Wordpool" Based on Narrowleaf milkweed Observations. Photograph by the Author.

based on your small corner noticings. Add, delete, and combine words to form a poem from your word collection.

Here is an example of a poem created from my wordpool:

Beyond the mist of morning,
a pattern of dots is seen.
Moving closer, I see a cloud of dark seeds
clinging to the stem of a narrowleaf milkweed.
The sight of Monarch butterflies echo through my memories
and outline my hope of seeing them here,
nourished by this native plant.

Though plant identification might begin as a solitary opportunity, sharing identifications through apps allows us to cultivate relationship skills both in the classroom and virtually. We can develop this SEL competency as we build confidence practicing communication skills about our plant identifications and observations.

Several years ago, I was lucky to attend a plant identification training that helped me better understand what it means to live in a biodiversity hotspot. Matthew Loftis, a Forest Aid Angeles Forest Coordinator, pointed out a small flowering plant, with five pink spoon-like petals. I was charmed by its delicate appearance, even more so when Loftis explained that its dark red pollen had been the mysterious cause of bees appearing with red legs the previous year. Not wanting to forget this diminutive beauty, I took a quick picture. I was eager to learn more about this plant, *Erodium cicutarium*, which I assumed was native to this chaparral region. Once I uploaded the image and read more about Erodium, I realized that it was one of the major invasive plant "offenders" in the Angeles National Forest. The workshop trainers explained that its aggressive growth and seed dispersal outcompeted native annual forbs for space. I now use iNaturalist as a layer of fact-checking as I learn more about forest flora and fauna in my small corner of the world and model this practice for my students.

Conclusion

A practice of tactile contact with the earth can foster kinship awareness, laying the groundwork for climate stewardship throughout one's life. Feeling connected enables climate stewards to persist, even when they are managing contradictory emotions or feeling anticipatory grief. Presenting varied ways for engaging in communal care will allow young climate stewards to keep going. This might look like slowing down before they pick up speed again.

References

Bratman, G., Anderson, C., Berman, M., Cochran, B., De Vries, S., Flanders, J., Folke, C., Frumkin, H., Gross, J., Hartig, T., Khan Jr, P., Kuo, M., Lawler, J., Levin, P., Lindahl, T., Meyer-Lindenberg, A., Mitchell, R., Ouyang, Z., Roe, J., . . . Daily, G. (2019). Nature and mental health: An ecosystem service perspective. *Science Advances, 5*, eaax0903. https://doi.org/10.1126/sciadv.aax0903

Doyle, B. (2019). *One long river of song*. Little, Brown.

Eisen-Markowitz, E., & Schragis, R. (n.d.). *What is this about?* Vent Diagrams. https://www.ventdiagrams.com/vision-and-values

Fish, K. (2020, March 9). The power of juxtaposition & creative alchemy: A microfiction prompt. *Flash Fiction Retreats*. https://flashfictionretreats.com/2020/03/09/the-power-of-juxtaposition-creative-alchemy-a-microfiction-prompt/

Great pacific garbage patch. (n.d.). *National Geographic*. Retrieved February 17, 2025, from https://education.nationalgeographic.org/resource/great-pacific-garbage-patch/

Harbage, C. (2018, November). *Butterfly preserve on the border threatened by Trump's wall*. NPR. https://www.npr.org/2018/11/01/660671247/butterfly-preserve-on-the-border-threatened-by-trumps-wall

Hickman, C., Marks, E., Pihkala, P., Clayton, S., Lewandowski, R. E., Mayall, E. E., Wray, B., Mellor, C., & Van Susteren, L. (2021). Climate anxiety in children and young people and their beliefs about government responses to climate change: A global survey. *Lancet Planet Health*, *5*(12), e863–e873. https://doi.org/10.1016/S2542-5196(21)00278-3. PMID: 34895496

Kroschel, W. (2019, August 22). *Shifting baseline syndrome*. Envirobites. https://envirobites.org/2019/08/22/shifting-baseline-syndrome/

Marchetti, A., & O'Dell, R. (2018). *Beyond literary analysis: Teaching students to write with passion and authority about any text*. Heinemann.

Mead-Brewer, K. C. (2018, March 8). *It's shaped like a grin, they say*. Cheap Pop. https://www.cheappoplit.com/home/2017/12/18/its-shaped-like-a-grin-they-say-kc-mead-brewer

Miller, B. (2013, November 16). We regret to inform you. *The Sun Magazine*. https://www.thesunmagazine.org/articles/21675-we-regret-to-inform-you

Miller, B., & Paola, S. (2012). *Tell it slant: Creating, refining, and publishing creative nonfiction*. McGraw-Hill.

Moore, L. (1985). *Self-help: Stories*. Knopf.

Niven, A. (2021, October 31). On Rannoch Moor. *Bath Flash Fiction Award*. https://www.bathflashfictionaward.com/2021/10/audrey-niven-october-2021-commended/

Plutchik, R. (1980). *Emotion: A psychoevolutionary synthesis*. Harper & Row.

Relles, D. (2020, February 18). Eulogy. *Passages North*. https://www.passagesnorth.com/issue-41/eulogy-by-dina-l-relles

Sprenger, M. (2020). *Social emotional learning and the brain: Strategies to help your students thrive*. ASCD.

Swan, H. (2020). *A kinship with ash*. Terrapin.

Thomas, J., Thomas, D., & Hazuka, T. (1992). *Flash fiction: Very short stories*. W.W. Norton & Company.

Thompson, C. E. (2022, July 6). *Laughter is the ultimate unifier. Can it work for climate action?* Grist. https://grist.org/fix/arts-culture/humor-comedy-engages-people-climate-science/

Wooldridge, S. G. (2009). *PoemCrazy: Freeing your life with words*. Crown.

Wray, B. (2023). *Generation dread: Finding purpose in an age of climate anxiety*. The Experiment, LLC.

7

Project-based Learning Design That Fosters Environmental Literacy

I've long been drawn towards project-based learning because I think it's so important that our learners enjoy opportunities to *chase questions*: open-ended questions that evolve or become refined as students deepen their understanding. Yet, I recognize how difficult it is for educators to design parameters for an ongoing project when we are trying to meet accountability targets and work within existing school schedules.

In earlier chapters, I've discussed resources that assist students in choosing their next steps as they narrow or broaden the scope of their investigation:

- Mentor texts
- Community mapping
- Hexagonal thinking
- Causal diagramming
- VUCA
- SDGs
- Power mapping

By connecting classroom activities with real-world problem-solving, we bring relevance to the learning experience. The challenges that plague my own PBL planning revolve around my facilitation and decision-making: What can we concretely do to help students dig into an inquiry experience that matters to them? How can we structure the "messy middle" of a project? How do we balance embracing open-ended investigation with scaffolding involving benchmarks and steps?

I've realized that connecting students with resources that pique their interest creates conditions for them to choose their own models and guides. It benefits students when we model how to find and select these resources.

In-person and Virtual Museum Visits

If we think of museums as a type of mentor text, both the selection of the museum objects and the staging of the museum objects offer "mentor moves" for us to consider. Museum storytelling stokes intergenerational awareness as a trusted source of information outside school walls that continually puts the past, present, and future in dialogue.

Offering opportunities that allow students to slow down and engage in slow looking can create receptivity to environmental calls to action that are otherwise ignored. In the attempt to create a sense of urgency for climate action, we might decide to subject our students to a parade of dire statistics. This onslaught of information can have the opposite effect: instead of moving students from inaction to action, we can inadvertently move them from inaction to despair. When aiming to equip students with the knowledge needed to take action, we can try to anticipate this common side effect of information glut. One way we can do this is to use visual texts as a hook.

Activity: Museum Exhibit as Mentor Text

Invite students to browse the *A Few Degrees More* online exhibit at the Leopold Museum in Vienna and to spend time gazing at the 15 paintings displayed. Through a collaboration with the Climate Change Centre Austria and renowned scientists from various fields, the museum exhibitors were able to determine the precise number of degrees by which climate change would impact the landscapes depicted in the painting and how it would affect the natural environments captured by the artists. Observing the tilted artwork literalizes the concept of "tipping point" in a way that provokes thoughtful scrutiny. A tipping point is the critical point in a situation, process, or system beyond which a significant and often unstoppable effect or change takes place.

Reflection:

> How does looking at the tilted paintings make an abstract concept more concrete?
> How does art make us more receptive to scientific warnings?

The exhibit offers a novel way to process the implications of data visualization. Consider pairing an exhibit painting with a different example of climate data visualization, such as the Keeling Curve. The Keeling Curve is a daily record of global atmospheric carbon dioxide concentration maintained by Scripps Institution of Oceanography at UC San Diego. If we study current Keeling Curve readings and these exhibit paintings together, what additional noticings might we surface?

Activity: Imagining a Museum Exhibit

In his 2016 book *Dive into Inquiry: Amplify Learning and Empower Student Voice,* Trevor Mackenzie describes four pillars of inquiry that ground classroom learning in intrinsic motivation:

- Explore a passion
- Aim for a goal
- Delve into your Curiosities
- Take on a New Challenge

TABLE 7.1 Brainstorming a Museum Exhibit Focus

Pillars of Inquiry	Brainstorming a Museum Exhibit Focus:
Explore a Passion	What is a museum topic you wish to share with the general public?
Aim for a Goal	What is an issue that deserves greater exposure?
Delve into your Curiosities	Which topic inspires you to deepen your knowledge about it?
Take on a New Challenge	How does a museum engage in problem-solving?

If we weave these pillars with museum exhibit design, students can launch their investigation and move in the direction they wish to chart:

Making space for "make-believe"—imagining a museum—is one way we can leverage curiosity about the world, potentially altering a research task from one of compliance to one of engagement. Invite students to create a proposal for a designed space that shares information about something they value.

Journal Prompt: You have the opportunity to imagine your own curio cabinet: an opportunity to mentally collect objects that represent topics that interest you. Take five minutes to describe three objects that hold personal value for you. Be sure to explain why.

It can benefit students to have opportunities to think about how we "stage" knowledge—how we present information and artifacts to elicit greater understanding about a topic deserving attention.

These questions can be used for either physical or virtual museum visits. For example, after visiting the Natural History Museum of Los Angeles County, my students began to envision a museum exhibit.

Prompts for Brainstorming your Museum Exhibit:

- What three objects will be the focus of your museum exhibit? Explain why.
- Exhibits are arranged to tell a story. How will your exhibit "tell a story" about its subject in a compelling manner? Describe how it does so.

- Did the museum have its guests walk through "linear" time (through chronological history)?
- Explain who your museum's target audience is and why.
- What features of the *Natural History Museum* will you "borrow" for your planned museum?

Giving students time to be reflective about the intentional design of museum spaces offers opportunities to think about the difference between individual and collective memories. It can develop awareness about how generational identity is powerfully informed by nostalgia and the passage of time. Comparing responses to museum objects illuminates how objects are imbued with different meanings depending on context and point of view.

Picture-book Inquiry

My high school students have fallen in love with "story time": picture books are passed out to student groups, and one member of each group reads the book aloud. By juxtaposing a picture book with a longer text, mental friction is produced. Side by side, the topics explored in each text bounce off each other and raise questions that would not be asked otherwise.

This train of thought leads me to think about the influence of picture books: how important it is to have access to stories at an early age, when we're just starting to make sense of our place in everything. As a model case study, I'll share steps for building literature circle discussions around picture books deliberately put into conversation with *Parable of the Sower*. This exercise helps my students take ownership of our class time, as their noticings and questions became the driver of our class discussions.

Selected Picture Books:

- *One Plastic Bag: Isatou Ceesay and the Recycling Women of the Gambia*
- *Outside In*
- *Because of an Acorn*

Step One: Getting To Know Their Picture Books

For the first read-through, the groups designated a member to do the read-aloud, who would make sure to display the pages to the group before turning each page. Then each member would process the picture book contents through a "head, heart, gut" activity:

- Head: What did the story make you think of?
- Heart: How did it make you feel? (Specifically, how did it do that? Through illustration style, repetition, memorable characters?)
- Gut: What resonates with you? (Something that resonates with you is similar to something you already think or believe.)

What became clear from listening to my students discuss their first impressions of their picture book was that thinking about it alongside Butler's novel allowed them to zoom in. Thrown against a new backdrop, important details of the novel became more noticeable. Through reading about Isatou Ceesay's effort to recycle the plastic bags that were polluting her community, one group better understood how Lauren's coming-of-age journey occurs as the disintegrating United States faces converging environmental and economic crises. The accumulating plastic bags reminded them of a striking metaphor from the novel: the "land sharks" waiting to find a way into the walled community. Both Isatou and Lauren are able to foresee and are willing to address problems that will only snowball.

Step Two: Exploring Picture-book Elements

For a second read-through, groups designated a different member to read the story and flip the pages while everyone looked carefully at the elements of a picture book that might be overlooked: its title, cover, endpapers, gutter, typography, color, and illustration style. After carefully considering these elements, I asked them to make connections with some aspect of Butler's novel. Then, each group wrote a bank of connections on a poster chart to be displayed throughout the rest of the unit.

While flipping the pages of *Outside In*, one student noticed that the illustrator uses warm and bright colors when describing the outdoors and darker colors when describing indoor scenes. He explained that these color associations usually are reversed in his head. His group made the connection that the walled community in Butler's novel is the so-called "safer" place compared to the outside. The illustrator's use of color helped my students deem the boundary between safe and threatening places in the novel to be illusory. Nevertheless, this group interestingly described "outside" as a type of ally in both books. The picture book's representation of "outside" connects it with its readers' increased social-emotional wellness and overlaps with Lauren's vision of a safer place up north where she can establish her Earthseed community.

Step Three: Documenting Emerging Questions

The ongoing usefulness of the text juxtaposition became clear once my students began to connect picture-book imagery and words with thematic ideas they had identified in the novel. The student group reading *Because of an Acorn* had identified change and growth as key thematic ideas that spoke both to Lauren's coming-of-age journey and the necessary mental adjustment the Robledo community needed to make in order to survive. Flipping through the picture book pages, my students observed how the illustrated interconnected layers of an ecosystem demonstrate so clearly the cause-and-effect dynamics that are always taking place. As a result of juxtaposing the texts, this student group generated these questions, which they used to lead discussion in subsequent classes.

- How does an acorn's life cycle relate to Lauren's life?
- How does the idea of a forest become similar to the change Lauren is trying to seed within her community?
- What role does change play in a young person's coming-of-age journey?

The visual stimuli enjoyed through these text juxtapositions helped my students more fully grasp concepts as they thought together. Picture-book inquiry fulfills the same function as other project starting points. When we begin a project, I create some sort of "entry event": via a video, a trip, or a story that can offer us a way to get oriented. A coordinate location from which to start as we get used to navigating with our own compass.

A Peek Into Their Practice

Kirstin Bullington is the Next Energy Engineering Instructor at the Richland Two Student Innovation Center in Columbia, South Carolina. Kirstin has been teaching for 22 years and currently works with students from five different high schools. She noticed students' interest and aptitude really explode when they have a real-world problem to solve.

"That's what I love about project-learning. You draw kids in with the authentic that may not have considered STEM careers. Sometimes your best problem-solvers are creative kids who aren't afraid to fail because they've done it before."

Kirstin loves seeing how students course-correct during a project. "The first idea failed. Where do we go from here?" Two students from her clean energy engineering class were studying the oriental hornet during a biomimicry design competition. The oriental hornet is able to capture sunlight and convert it to energy via its outer layer. The students attempted to mimic the hornet's outer layer by placing filters on solar panels but were unable to simulate the same process. They then took a closer look at the shingle pattern on the hornet's back, which inspired them to use a 3D printer to add panels to reflect additional UV light onto a solar panel. Their design for the new prototype was awarded first place.

"The best projects are projects where there is a real problem to solve." Kirstin's former colleague is now the head of a rural school in Senegal. The school lacked electricity to power a donated photocopier, and she wondered if they could work together on a power solution. They knew the criteria: they needed something to be made from as many local materials as possible. They knew the constraints: they needed it to be built and repaired by the students themselves at the school and without electricity. The students designed simple solar boxes with a modular power system to create electricity and shipped them over. The design went on to earn a U.S. patent.

"We throw kids into projects, but we don't teach them how to manage their work and manage others. We all need more experience developing project management skills." Kirstin has found the resources provided by the Project Management Institute Educational Foundation to be useful aids. "I do a lot of scaffolding in the beginning of our first course, but then I scale it back. Students need structure to effectively manage a timeline and goals."

Kirstin sheds light on one reason many teachers are intimidated by the thoughts of project-based learning. We can doubt our students' abilities to manage a schedule, stick to deadlines, and develop a scope of work when the direction of the project seems amorphous. Helping students develop norms for collaboration at an early stage of the project can minimize risks that arise when the line of communication has broken down.

From the outset, communicating a clear vision about project goals gives scheduling a purpose. Articulating priorities allows group members to devote their energy to tasks that deserve more attention and care. Stakeholders can have an outsized influence on decision-making. It is a good idea to have clarity about where pressure might come from and how to handle communication with influential stakeholders. Being realistic about timelines means making a point of formal check-ins, so there is already an established communication line when curveballs happen.

The possibility of strengthening or developing new networks of support can be a wonderful byproduct of working on a long-term project. Dayna Laur and Jill Ackers (2017) encourage educators to create space for mentors and community partners. These partners can provide feedback during the project, not just at the

TABLE 7.2 Norms for Collaboration Planning Sheet

Goals: What are we hoping to accomplish?
Priorities: What do we need to prioritize in the order of tasks we set out to complete?
Stakeholders: Who influences our goals and planning?
Timetable: How often will we check in and formally assess our progress?
Partners: Who can support our goals and amplify the impact of our project?

end while beholding a final product. The real-world challenges our students address require comfort with tapping into community resources.

This email message offers an example of reaching out to a community expert. I share it with my students as an example of how to start conversations beyond classroom walls.

> Dear Dr. Wilson,
>
> I am a high school teacher who recently discovered some of the work you do as a Green Infrastructure Project Manager. I watched a video about the Native Garden Kits you distributed, which aims to help homeowners transform their lawns to southern California native plants. Your efforts inspire me as I share a commitment to helping pollinators find habitat and food sources in local ecosystems. I am contacting you because I was wondering if you would be able to answer a few questions I have regarding the important work you do:
>
> As you worked on this project, did your organization partner with other organizations to build capacity for this project? To increase awareness?
>
> When you think about community impressions of native plants, do you notice trends in what people tend to overlook or simplify when discussing this topic?
>
> As a teacher, I am trying to figure out the best way to communicate the importance of native plant gardening. What do

you think is the root cause of so much non-native planting in yards and public gardens?

Your responses will help me develop a plan to increase awareness of this issue and recommend actions that can be taken. Thank you in advance for any insight you can provide.

Sincerely,
Xochitl Bentley

Conclusion

This book was written with the needs of classroom educators in mind—the lived experiences of teaching practitioners who are searching for accessible entry points and pathways to cultivating students as climate stewards. With these entry points, we can eliminate barriers to participating in solutions and support students as they create their own multiplying effect through storytelling. Every school can be a "climate victory garden." Inspired by World War-era victory gardens, climate victory gardens aim to maximize soil protection and minimize climate impacts beyond the boundaries of the garden. Our storytelling skills can help us develop the root system we need so we can become the ancestors our descendants deserve.

References

Butler, O. (2019). *Parable of the sower*. Grand Central Publishing. (Original work published 1993)

Laur, D., & Ackers, J. (2017). *Developing natural curiosity through project-based learning: Five strategies for the PreK-3 classroom*. Routledge.

Mackenzie, T. (2016). *Dive into inquiry: Amplify learning and empower student voice*. EdTechTeam Press.

Paul, M., & Zunon, E. (2015). *One plastic bag: Isatou Ceesay and the recycling women of the Gambia*. Millbrook Press.

Schaefer, L. M., Schaefer, A., & Preston-Gannon, F. (2016). *Because of an acorn*. Chronicle Books.

Underwood, D., & Derby, C. (2020). *Outside in*. Clarion Books.

For Product Safety Concerns and Information please contact our EU
representative GPSR@taylorandfrancis.com
Taylor & Francis Verlag GmbH, Kaufingerstraße 24, 80331 München, Germany